BIGFOOT
FILM JOURNAL

The Moment by Michael Rugg. The scene depicts the moment Patterson and Gimlin spotted the creature on the north side of Bluff Creek, California, on October 20, 1967.

BIGFOOT
FILM JOURNAL

Christopher L. Murphy

Edited by Roger Knights

Associated Editors/Consultants:
Thomas Steenburg
Daniel Perez

hancock house

ISBN 978-0-88839-658-7
Copyright © 2008 Christopher L. Murphy / Third printing 2010

Cataloging in Publication Data

Murphy, Christopher L. (Christopher Leo), 1941–
   Bigfoot film journal: a detailed account & analysis of the Patterson/Gimlin film circumstances & aftermath / Christopher L. Murphy.

   Includes bibliographical references and index.
   Also available in electronic format.
   ISBN 978-0-88839-658-7 (bound)

   1. Sasquatch—California—Humboldt County. I. Title.

QL89.2.S2M868 2008a   001.944   C2008-903263-2

All rights reserved. No part of this publication may be reproduced, stored in a retrieval system or transmitted, except for personal use, in any form or by any means, electronic, mechanical, photocopying, recording, or otherwise, without the prior written permission of Hancock House Publishers Ltd.

Editor: Roger Knights
Associated Editors/Consultants: Thomas Steenburg and Daniel Perez
Book and Cover Designs: Christopher L. Murphy
Front Cover Image: Patterson/Gimlin film
Back Cover Images: Film site model by C. Murphy; Sasquatch artwork by Brenden Bannon, based on frame 364 of the Patterson Gimlin film; Casts from the Patterson/Gimlin film site.

We acknowledge the financial support of the Government of Canada through the Book Publishing Industry Development Program (BPIDP) for our publishing activities.

Printed in the USA

Published simultaneously in Canada and the United States by:

HANCOCK HOUSE PUBLISHERS LTD.
19313 Zero Avenue, Surrey, B.C. Canada V3Z 9R9

(604) 538-1114 Fax (604) 538-2262

HANCOCK HOUSE PUBLISHERS
#104-4550 Birch-Bay Lynden Rd, Blaine, WA U.S.A. 98230
(800) 938-1114 Fax (800) 983-2262

Website: www.hancockhouse.com
Email: sales@hancockhouse.com

Contents

Acknowledgements	6
Foreword	7
1. Background and Insights	8
2. Introduction	9
3. Prelude to the Encounter	12
4. The Encounter	18
5. Crucial Decisions	34
6. Credibility of the Source	48
7. The Scientific Community and the Press	59
8. The Canadian Controversy	67
9. Patterson and Gimlin in Hollywood and New York	72
10. European Scientific Reviews	80
11. Current Findings and Insights	83
12. The Major Issues	86
13. The Major Hoax Claims	89
14. Taking Stock	97
Bibliography	99
Photograph Credits/Copyrights	100
General Index	103

Composition Note: *Apart from direct quotations, I have used the term "bigfoot" rather than "sasquatch" throughout this entire presentation. In this case, use of the latter would not be appropriate. Furthermore, I have chosen not to capitalize the word and to use it as both a singular and plural term.*

Acknowledgements

First and foremost I acknowledge and thank Robert Gimlin and Patricia Patterson for their warm friendship and willingness to provide information on the intriguing movie we call the Patterson/Gimlin film.

Additionally, much of what I present is based on the work of individuals who have put forth significant effort in examining the film circumstances and its aftermath. My gratitude is extended to:

> René Dahinden+
> John Green
> Peter Byrne
> Daniel Perez
> Thomas Steenburg
> Roger Knights
> Clifford Crook
> Barbara Wasson Butler+

There are also those who applied their various skills to the analysis and interpretation of the film subject herself, providing a clearer vision and understanding of what we believe was probably a natural bigfoot. In this connection, I thankfully acknowledge:

> Dr. Grover S. Krantz+
> Dr. Jeffrey Meldrum
> Dr. Henner Fahrenbach
> Jeff Glickman
> Dmitri Bayanov
> Igor Bourtsev
> Doug Hajicek
> Marlon Davis
> Richard Noll
> Brenden Bannon
> Peter Travers
> Yvon Leclerc
> RobRoy Menzies

Furthermore, there are those who have provided me with knowledge in all aspects of the overall bigfoot phenomenon, which has benefited me throughout this work. I acknowledge with appreciation:

> Bob Titmus+
> David Hancock
> Ray Crowe
> Robert Alley
> Thom Powell
> Al Berry
> Robert Morgan
> Joedy Cook
> Larry Lund
> Loren Coleman
> Bobbie Short

And credit where credit is due, I acknowledge Greg Long, who despite his conclusions did bring to light a lot of valid information on the filming circumstances.

Note: "+" indicates deceased.

Foreword

In 2004, Hancock House Publishers reprinted Roger Patterson's book, *Do Abominable Snowmen of America Really Exist?* Glen Koelling, through Franklin Press, published this book in 1966. It was reprinted twice after that year, and finally by Hancock House (2005) under the new title, *The Bigfoot Film Controversy*. The Hancock House version includes the complete Patterson work verbatim, along with a supplement that I was invited to write.

My supplement provides an overview of the filming story, twelve film-frame images, scientific reports on the film, and a section in which I address the allegation by Greg Long that Roger Patterson and Bob Gimlin fabricated the film.

The overview of the filming story I provide is essentially an updated version of that which I wrote to accompany a report prepared by Jeff Glickman of the North American Science Institute in 1998 entitled, *Toward a Resolution of the Bigfoot Phenomenon*. Unfortunately, a book that was to contain this material was withdrawn from publication.

The section addressing Long's allegation is my own work plus that of several other researchers, particularly John Green and Richard Noll.

If one wishes to get a good appreciation of the material I have mentioned, then I highly recommend *The Bigfoot Film Controversy*.

In the course of events over many years, I have been of the opinion that we (the bigfoot community) need to put more on the table concerning the remarkable Patterson/Gimlin film. In other words, to move in closer and present and examine all of the circumstances and issues surrounding the film. In this way, we might be able to eliminate (or at least minimize) all of the criticism and skepticism that somehow evolves in the minds of journalists and others about the film.

Here then, is everything I have gathered, neatly arranged and discussed, topic by topic.

Sasquatch/Bigfoot "Homeland"

The Patterson/Gimlin film was taken in a part of the world that has massive unexplored regions. Such are situated in Alaska, the Yukon, British Columbia, Washington, Oregon, and California, as seen here without north/south borders. If bigfoot migrated from Asia, as we suspect, this entire area was likely its original "homeland," and it is here that about one-half all North American bigfoot sightings/incidents take place. Certainly, if anyone were to get film footage of the creature, then the "homeland" would be the best bet. By the same token, of course, such would aid in giving "credibility" to a hoax.

1. Background and Insights

My entrance, as it were, into the field of bigfoot studies occurred in June 1993. My son Daniel was doing a college paper on the subject, and he asked me to accompany him to interview René Dahinden. Prior to that time I had seen the Patterson/Gimlin film on television once or twice, but gave it little or no credence. I was, however, aware that someone (Rant Mullens) had hoaxed bigfoot footprints. That was the extent of my knowledge of the subject.

The interview with René went well. He showed us photographs and casts that heightened my interest in the subject. Further visits with René followed, during which he showed us the actual Patterson/Gimlin film. In the meantime, I read René's book, *Sasquatch/Bigfoot,* and books by John Green. My son and I saw a potential in marketing footprint casts along with posters of the creature made from film frames. René agreed to give it a try, and these items were made available to the general public through my company, Progressive Research. Next came republication of Roger Patterson's book, *Do Abominable Snowmen of American Really Exist?,* and Fred Beck's book, *I Fought the Apemen of Mt. St. Helens,* through my company, Pyramid Publications.

During my numerous visits with René, we had long discussions on the Patterson/Gimlin film, which naturally led into the *circumstances* related to the film. I often reflected on John's Green's statement that if just one bigfoot sighting could be fully substantiated, then bigfoot is real. The same holds true in reverse for the film. If just one piece of evidence related to the film *proves* the film is a hoax, then the film is a hoax. It would not matter how real the creature appears or how it could possibly have been created. The film itself has certainly withstood detailed analysis and is difficult to dismiss as a hoax. The film circumstances, however, have not been subjected to the same level of scrutiny.

I started documenting this work in 1996, working directly with René. At the time, our objective was to produce a book on the film, tentatively titled *Bigfoot at Bluff Creek*. Several installments were provided to René, and he was impressed with the method (explained in the next section) I used to organize the information. Unfortunately, René and I parted company before the work was completed. I therefore continued research on my own. I retitled the work *Circumstantial Evidence – The Patterson & Gimlin Film.* In 1999, I sent a draft manuscript to John Green and Dr. Henner Fahrenbach, who commented favorably. While I provided a few copies to other researchers, I did not publish the work because I was not comfortable with some of the information, and because the work lacked proper film-frame illustrations. I have since reasonably satisfied myself on the information issues, and my other works published by Hancock House have resolved the need for film illustrations.

My personal association with other people directly involved with the film, such as John Green, Bob Gimlin, and Patricia Patterson, together with high profile researchers, such as Dmitri Bayanov, Daniel Perez, and Thomas Steenburg, has provided me with knowledge not generally known. However, I do not imply that they agree with all of my assumptions and speculations.

René (right) and me in 1993. During the time I worked with René, hardly a day went by that we did not discuss new sightings and other bigfoot incidents. (I met with him two or three times a week.)

René lived and worked at the Vancouver Gun Club in Richmond, B.C. He is seen here with my two sons, Chris and Dan. The setting was perfect for discussing bigfoot, and I have many fond memories of my visits.

René had a great sense of humor and was a lot of fun to work with. He passed away on April 18, 2001. He would have been 71 in August of that year.

2. Introduction
Coming to grips with reality

The story of the Patterson/Gimlin film is complicated, confusing, and lacking critical details. To make matters worse, as the years roll by, memories fade and important facts become harder and harder to obtain. Indeed, in most published works that mention the film, the authors virtually tiptoe around the film's history. They have done this to avoid becoming embroiled in the mass of loose ends and dead ends.

There are only two people who know (or knew) all or most of the historical answers—Roger Patterson and Bob Gimlin. Patterson, of course, is no longer with us; he passed away in 1972. There are many questions he should have been asked before he passed away. Investigators failed to realize the importance of certain details. We are, therefore, left with Bob Gimlin alone. Bob has certainly done his best to answer questions. Thomas Steenburg and I recently (2003) spent some five hours with him and went through the events of the filming and its aftermath in considerable detail. The information Bob provided was straightforward and simple. There are some specific things he just does not know or cannot remember, and he told us exactly that when such was the case. I am totally convinced beyond any doubt that he is not hiding anything. Certainly, we would be much further ahead if questions asked of Gimlin in recent times had been asked thirty-nine years ago, but there is nothing we can do about that now.

There is one other major player who I thought for many years could *possibly* shed a little more light: Patterson's brother-in-law, Al DeAtley Jr. I was told that DeAtley gave the cold shoulder to researchers, so I never attempted to interview him. However, current research by others indicates that DeAtley remembers very little of what happened on October 20 to 22, 1967. While this might appear odd on the surface, trying to remember details of an event (even a major event) that occurred 40 years ago is impossible for many people, including myself.

An obvious question that comes up is, why did the major bigfoot investigators at the time fail to get specific answers and tangible evidence related to the history of the film? To begin, they thought the search was effectively over. In other words, that it would be only a matter of a few months before a sasquatch was captured or killed. When this did not occur, they concentrated on proving (or disproving) the *direct* authenticity of the creature filmed. They considered only the major factors—Patterson and Gimlin's personal credi-

This photograph of your author was taken by Gerry Matthews when he came to interview me for a presentation on his website (www.westcoastsasquatch.com). Since becoming involved in the bigfoot issue, my life situation allowed me to spend a tremendous amount of time researching, studying, and writing information on the subject. In short, my time was, and is, my own. Although there are certainly some questions related to the Patterson/Gimlin film for which I do not have answers, I don't think the information went to the grave with either Roger Patterson or René Dahinden. Nor do I think it is being withheld as some deep, dark secret by those people directly associated with Roger Patterson.

From left to right, Bob Gimlin, John Green, Thomas Steenburg, and Dmitri Bayanov. I took this photo on our way to the Willow Creek, California, symposium in 2003. Bob traveled with Tom and me part of the time, which provided a great opportunity to discuss the film and enjoy the great countryside.

bility and film details (as opposed to *filming* details).

We must remember, the investigators were not financed. They had limited time and resources. Furthermore, some of the questions that needed to be asked were not those a person would commonly ask in the investigators' situation. They were not policeman. They could not demand information. Finally, some of the questions have *evolved* as a result of subsequent findings. They are not questions one would have thought about at an earlier time.

Although we do not have all the answers, we do have a considerable amount of *circumstantial evidence* that points to many answers. Furthermore, we can also make *reasonable* deductions from various known factors to give us yet a clearer picture of what actually happened. When we sort out and analyze the information we have, it is surprising how it *truly supports the authenticity of the bigfoot creature filmed.*

In this work, I make an attempt to fully examine the *circumstantial evidence*. Although a lot of what I present can be backed up with hard evidence (documents, photographs), most information is anecdotal—based on what people have simply said or written. What people simply say or write is derived from what they perceive or what they have gathered from other people. There is often nothing to prove that their perception or secondhand information is correct. In this connection, we generally assign credibility to the information on the basis of the person's education, occupation, or station in life. We usually give statements from people with "credentials" more credibility. I believe that if two half-blind eminent anthropologists were to see a bigfoot, such would be given more credibility than 100 sightings made by non-scientific people with perfect vision—perhaps even more than that given the Patterson/Gimlin film.

As a result of these conditions, one might suppose that the best thing to do is refrain from writing about anything for which hard evidence is lacking. This is especially applicable when we write about people, because it is unfair to make an assessment of a person based on what might only be gossip. Nevertheless, as the Patterson/Gimlin film is so important, it is essential that we consider everything concerning the film to formulate opinions on its authenticity. Moreover, very often things said and written without hard evidence result in such evidence coming to the surface.

In this work I have pieced a lot of *possible* facts together and then analyzed the results, much like mixing various chemicals to see what happens—a dangerous practice. Certainly, I expect some mixtures will explode. I acknowledge that there may be others who

> "Although a lot of what I present can be backed up with hard evidence (documents, photographs), most information is anecdotal—based on what people have simply said or written."

The amount of literature that has been written on the bigfoot issue is quite remarkable. I believe I have the most important works, but certainly not all of the books available. On the right in this photo are the two books that categorically dispute either the Patterson/Gimlin film or the creature in general. Surprisingly, not one of these books, pro or con, provides a scientifically acceptable explanation for the bigfoot phenomenon. Of course, as soon as such an explanation is provided, then the trail of books advocating or disputing bigfoot's reality will end.

know *some* of the "chemicals," as it were, better than I, but unfortunately the sharing of *full* knowledge in the bigfoot field leaves a lot to be desired. If I am wrong in my own *perceptions* then so be it—every writer reserves the right to be wrong.

In preparing this work, I had to consider the *nature* of the overall story. To relate the story in conventional format borders on the impossible. In order to fully understand what is being presented, the reader would have to be taken on numerous tangents. I have therefore utilized, where applicable, a referencing system. I present the general chapter information in concise story format without a lot of details, and virtually no analysis. I then proceed to provide specific details and analysis in expanded footnote format **immediately after the story section.** Where a footnote number is shown in the story section, that number references the appropriate point (same number) in the Details and Analysis section.

I provide both sides of the arguments—information supporting the film's authenticity and that which questions it. I also look at the aftermath of the film and its impact (or lack of same) on the scientific community.

Everything I present is based on the knowledge I *currently* have. Since I started documenting information, I have had to continually revise my findings in view of new revelations—basically old buried information that found its way to the surface. Certainly, new information will continue to come to light, so this work will never really be finished.

This is the coil-bound manuscript I completed in 1999. I sent it to two researchers and provided a computer file to a few others. I then basically buried it for the reasons I have stated. I was much more negative in those days, but still came out on the high "plus" side as to the film's credibility. What I have learned in the past seven years has filled in a lot of gaps. Had this work been available at that time, it might have tempered some of the skepticism of the two recent film "debunkers." But then again, I think their minds were made up before they started.

"I provide both sides of the arguments—information supporting the film's authenticity, and that which questions it."

3. Prelude to the Encounter

Roger Patterson[1] had been interested in the bigfoot phenomenon since the early 1960s and had seen alleged bigfoot footprints firsthand. He and his friend Bob Gimlin[2] had gone into different wilderness areas in Washington State many times to search for the creature. Often, they would be following up reports of sightings or footprints. Highly influenced by the writings of Ivan T. Sanderson, a zoologist, Patterson compiled his own book entitled, *Do Abominable Snowmen of America Really Exist?*[3] The book was published in 1966.

At some point in 1965 or 1966, Patterson decided to make a film documentary on bigfoot. He subsequently rented a movie camera[4] in May 1967 for this purpose. He wished to find and film fresh footprints as evidence of the creature's existence.

During late August and early September 1967, Patterson and Gimlin were exploring the Mt. St. Helens area. While they were away, Al Hodgson and the late Syl McCoy, friends in Willow Creek, California, telephoned Patterson's home (September 5, 1967) to report that large human-like footprints had been found along roads being constructed on Blue Creek Mountain, Six Rivers National Forest, California.[5] Patricia Patterson, Roger's wife, took the message and gave it to her husband when he returned. This same area had been the scene of considerable bigfoot activity nine years earlier. It was here in 1958 that Jerry Crew,[6] a road construction worker, found large human-like footprints near Bluff Creek.[7] A subsequent press release on Crew's find made the word "bigfoot" the American name for the creature. Crew worked for Ray Wallace[8] a road construction subcontractor building roads in the area at that time.

Upon receiving the message from his wife, Patterson contacted Gimlin and the two made plans to investigate the Blue Creek Mountain findings. Patterson purchased (or already had) a number of 100-foot color film rolls[9] for the expedition.

Patterson and Gimlin traveled to the Bluff Creek area in Bob Gimlin's truck, taking with them three horses. We believe they left on October 1, 1967.[10] By the time the men arrived and set up their camp (at a place near Louse Camp, where Notice Creek and Bluff Creek join), road construction workers had all but destroyed the prints reported by Hodgson and McCoy. Patterson and Gimlin thereupon set out on horseback to explore the area. Patterson was intrigued with the scenery and autumn colors. He used 76 feet of his first film roll for general filming, including shots of himself, Gimlin, and

John Green is seen here measuring the footprints that had been found on Blue Creek Mountain. He and René Dahinden investigated the prints and called in Don Abbott of the British Columbia Provincial Museum. Abbot went to the site and saw the prints firsthand. (Photo, J. Green)

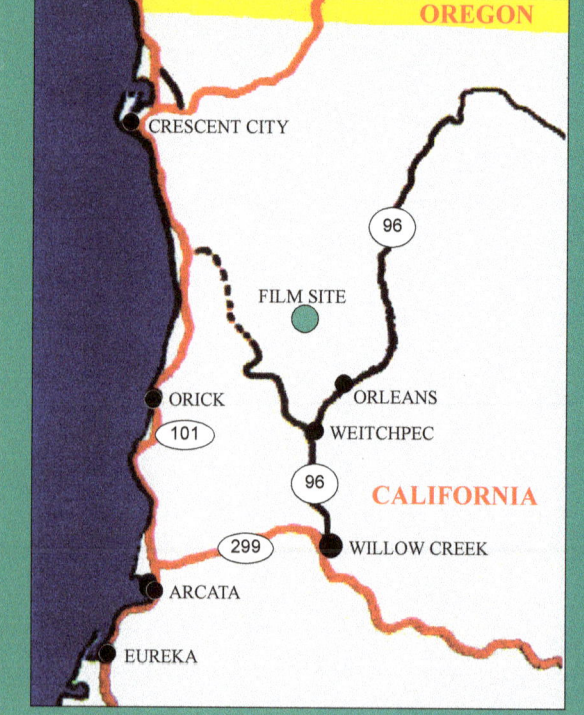

Northwestern California with the film site indicated.

the horses. After dark, the men drove through the area, where accessible, in hopes of a possible night sighting of a bigfoot.

Patterson and Gimlin were observed in the area twice by two members of a road building crew, Mr. and Mrs. Charlie Hooker of Oroville California.[11]

DETAILS AND ANALYSIS

1. Roger Patterson

Patterson was born in 1933 in Walls, South Dakota. His main source of income came from rodeo work. He married Patricia Mondor in 1956; the couple had three children: Sabrena, Clint, and Brad. At the time of the expedition, he lived about 7 miles southwest of Tampico, Yakima County, Washington. Patterson died of Hodgkin's disease on January 15, 1972.

2. Robert Gimlin

Gimlin was born in 1931 in Missouri. At the time of the expedition, he lived in the Union Gap neighborhood of Yakima, Washington. He now lives on the southwest outskirts of that city. He is one-quarter First Nations heritage and made his living as a truck driver. He had a side interest in training horses. Now retired, Bob's main interest is horse training. Gimlin is an experienced animal tracker, and it was this ability that attracted Patterson. He needed Gimlin to help him follow possible bigfoot tracks when such were found or reported.

3. Patterson's Book

Patterson's firm belief in the existence of bigfoot is evidenced by his book, *Do Abominable Snowmen of America Really Exist?* It is very likely that his belief enabled him to remain reasonably calm and in control when he saw the creature. In other words, he was psychologically prepared for an encounter. As a result, he was able to get the movie footage. Other people who report bigfoot sightings often state that they are so surprised they lose control. They fail to get photographs even though they have a camera with them.

4. The Camera

The camera Patterson rented was a standard Cine-Kodak K-100 16mm movie camera. This type of camera was marketed between 1955 and 1964. In size, it was about 9"x5.75"x2.5". It weighed about 5.75 pounds. There are some unusual circumstances regarding the camera itself. Patterson had rented it from Sheppard's Drive-In Camera Store in Yakima on or about May 13, 1967, and he was extensively overdue in returning the unit when he went to Bluff Creek. A war-

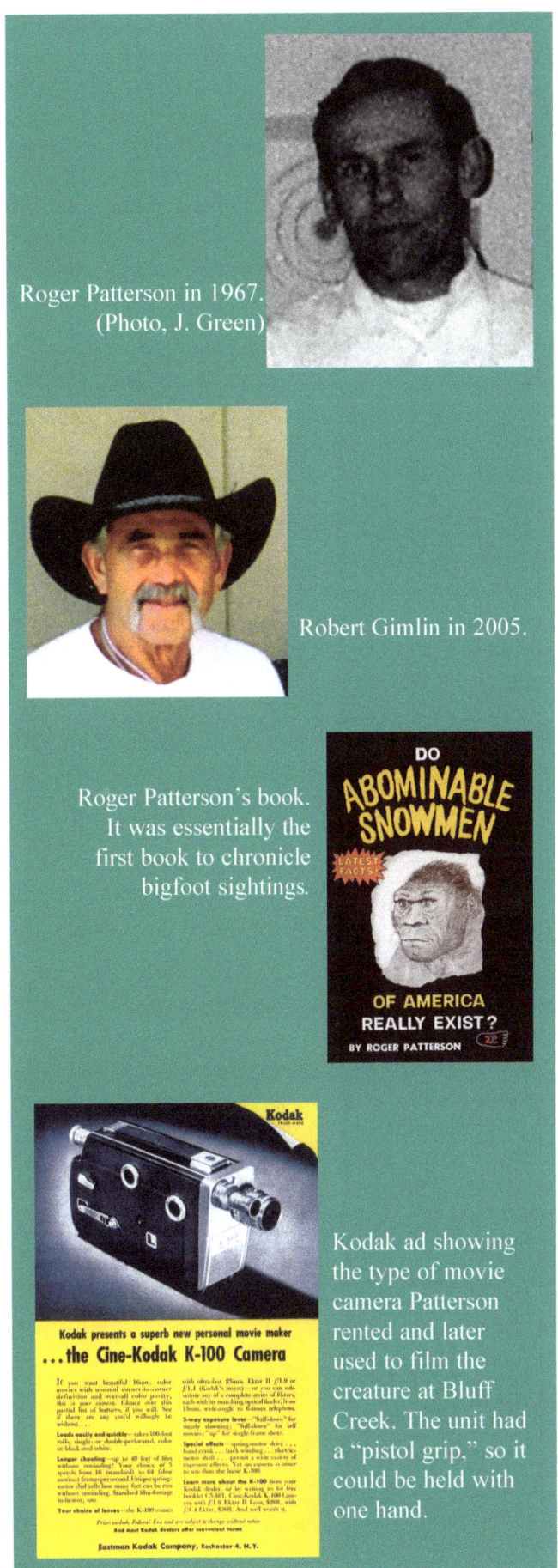

Roger Patterson in 1967. (Photo, J. Green)

Robert Gimlin in 2005.

Roger Patterson's book. It was essentially the first book to chronicle bigfoot sightings.

Kodak ad showing the type of movie camera Patterson rented and later used to film the creature at Bluff Creek. The unit had a "pistol grip," so it could be held with one hand.

rant was issued on October 17, 1967 to arrest Patterson on the charge of grand larceny for his failure to return the camera on time. This indicates that Patterson did not really know how long he would need the unit. Had he been party to a planned "fabrication" for his documentary, he would have known the exact time period for which he needed the camera and would have probably returned the unit on time.

With regard to the warrant, Patterson was taken into custody on November 28, 1967. At a court hearing the following day, he was released on his own recognizance. The charge was dismissed at a hearing on December 8, 1969. One final note: Bob Gimlin had been of the opinion that Patterson owned the camera.

While some people might "make hay" over the camera incident (i.e., cite it as evidence of Patterson's dishonesty), I really don't think this reasoning is either rational or sensible. Yes, he should have returned the unit on time and was irresponsible for not doing so. Regardless, the incident actually supports the film's authenticity. Surely, if he were involved in a hoax he would not want to provide a reason to doubt his honesty.

If the question is, do I think Roger Patterson really thought he could get a movie of a bigfoot creature? The answer is no. But Roger was a very clever person, and I think he reasoned that his chances of getting convincing evidence of a bigfoot were many times better with a movie camera than with a regular camera. Think about it. Where speed is essential, a movie camera snapping 16 pictures a second was far superior to single shot cameras available at the time.

5. The Prints on Blue Creek Mountain
The prints reported by Hodgson and McCoy had previously been reported (late August 1967) to John Green and René Dahinden, who had already gone to the area and investigated the findings. They brought in Don Abbott of the British Columbia Provincial Museum, and the road contractor actually held up the job while the researchers investigated the prints.

6. Jerry Crew (d. 1993)
A road was constructed into the Bluff Creek region starting in 1957, opening the area, which up to that time had been remote wilderness. On August 27, 1958, Jerry Crew, a road construction worker, noticed large humanlike footprints in the soft soil. Crew had heard of similar findings by a road gang about one year earlier at a location eight miles north. He showed the prints to his fellow workers, some of whom said they had also seen prints in the area. He saw additional prints about one

Jerry Crew in 1958.
(Photo, *Humboldt Times*)

Copy of the cast made by Jerry Crew in 1958.

month later and more on October 2, 1958. This time, he made a plaster cast of one of the prints and reported his find to the press.

Together with the footprint finds, there were also alleged bigfoot sightings and other unusual incidents in the area. Subsequent investigations at that time revealed tracks of six different sizes, indicating that a number of bigfoot frequented the area. Footprint sizes ranged from 12.25 inches to 17 inches long. These facts had made the Bluff Creek area a prime location for a possible bigfoot sighting, and it is certainly not unusual that footprints would subsequently be found in the same general area in September 1967.

7. Bluff Creek

We can certainly reason that the Bluff Creek area was a prime location for a possible major sighting. However, by the same token, we cannot discount that it was also a prime location to perpetrate a hoax. Some people even contend that the name "Bluff Creek" may be a subliminal message in this regard. I hardly think, however, that this reasoning has any significant validity.

8. Ray Wallace (d. 2003)

Wallace was a subcontractor to the firm Block and Company. He originally thought that the "bigfoot" incidents were the work of someone trying to disrupt his operations. He was very angry and brought in a man, Ray Kerr, to help catch the culprit. We are told Kerr and his friend, Bob Breazele, sighted a bigfoot while driving down the new road in mid October 1958.

It appears the publicity on Jerry Crew's findings, and subsequent public interest in bigfoot, prompted Wallace to think of ways in which he could benefit from the phenomenon (i.e., get personal publicity and generally have some fun). He thereupon fashioned a pair of large wooden feet and, *we are told*, made footprints in various Bluff Creek areas.

In November 1970, Wallace claimed that he had filmed a bigfoot creature in 1958 (my reference states 1957, but I am sure 1958 is correct). To my knowledge, nobody has ever seen this footage. Furthermore, according to Wallace, he said that he told Roger Patterson that Bluff Creek (the subsequent film site area) was the best place to go to possibly film a bigfoot. In Wallace's own words:

> I felt sorry for Roger Patterson. He told me he had cancer of the lymph glands and he was desperately broke and he wanted to try get something where he could have a little income. Well, he went down there just exactly where I told him. I told him, "You go down there and hang around on that bank. Stay up there and watch that spot." I told him where the trail was that went down to where that big rock was. I told

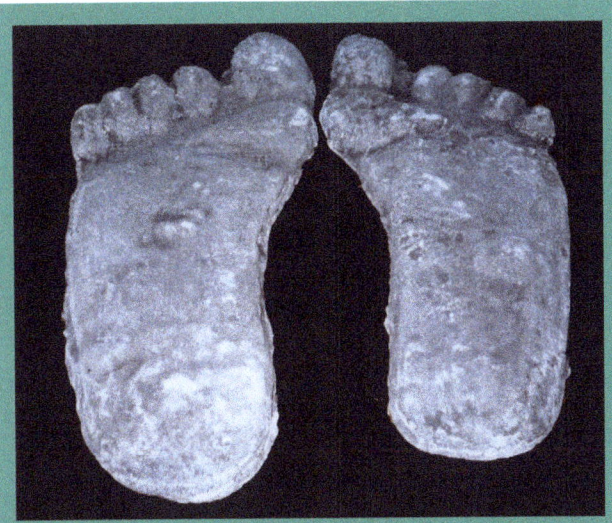

One example of casts from footprints made with wooden feet fashioned by Ray Wallace. He made many wooden feet of different sizes and shapes. Just when he started making the carvings is not known. He made one set that does resemble the prints found by Bob Titmus (1958), and one of the two 1967 Blue Creek Mountain print types. None of his "models" match the prints found by Jerry Crew (1958), or those left by the creature filmed by Patterson and Gimlin (1967). Despite a $100,000 reward offered by the Willow Creek–China Flat Museum after Wallace died for proof of how prints found are fabricated, Wallace's sons have not come forward. All they need do is demonstrate the process according to the guidelines to claim the prize. The last I heard is that the Wallace wooden feet are "family heirlooms."

This recent satellite photo shows the roads (as defined by Google Earth) that are presently in the Bluff Creek region. I do not know the specific road on which prints were found by Jerry Crew in 1958, however, we can see that all roads are very remote. Photo: Image from Google Earth © 2008 Tele Atlas; DigitalGlobe.

him where he could get those pictures down there, Bluff Creek.

The reference to "cancer" in Wallace's statement is interesting as Patterson was not diagnosed with "cancer" until 1970. Actually, he had Hodgkin's disease, which Wallace probably understood as "cancer." We also have a claim by Wallace that he knew who was in the "Patterson suit." Wallace would not provide the name, but stated that the person was a big Yakima Indian. Moreover, in or about the year 1972, according to John Napier (*Bigfoot,* p.87*),* Wallace was *said* to have 15,000 feet of color film of bigfoot. I think we can surmise that Wallace himself let this information be known.

On March 30, 1998, a Washington newspaper carried the following article concerning Wallace:

> A Toledo, Wash., construction tycoon by the name of Ray Wallace is offering a $1 million reward to the first person who can bring him a live bigfoot baby. The 79-year-old Wallace has spent more than 40 years tracking down the elusive Northwestern creature known as Sasquatch. If anyone does respond to Wallace's offer, he plans to raise baby bigfoot with "care and respect" and would like to train the creature to ride around with him in his pick-up truck and help with chores around his ranch.

Upon Wallace's death in 2002, his family stated to the press that he was responsible for starting the bigfoot issue. In other words, they stated that he fabricated the footprints found by Jerry Crew in 1958.

There are other stories concerning Ray Wallace, one about a baby bigfoot he claimed to have caught and attempted to sell for $25,000, asking for payment prior to delivery. When prospective purchasers (Tom Slick and Peter Byrne) asked to see the creature before making payment, Wallace told them it had become ill, so he let it go. Another concerns film footage he allegedly took of a bigfoot that he also tried to sell. Again, I believe he asked for advance payment. This second extortion attempt also failed. It is very difficult to sort out and make sense of anything concerning Ray Wallace. He was just a silly old man having fun causing confusion. His main "dupes" were newspaper reporters.

Unfortunately, the media was either unaware of, or chose to ignore, the background information I have presented. We may conclude beyond a doubt that Ray Wallace's credibility was highly questionable. He was known for circulating dubious bigfoot photographs, footprints, and tape recordings. Nevertheless, the fact that Wallace was a "man of means," has raised speculation that he could have financed a hoax relative to the Patterson/Gimlin film. Personally, I find this speculation absurd.

9. The Film Rolls — Type and Quantity
The film used was Kodachrome II daylight, color reversal movie film. This film was available in 50-foot or 100-foot rolls.

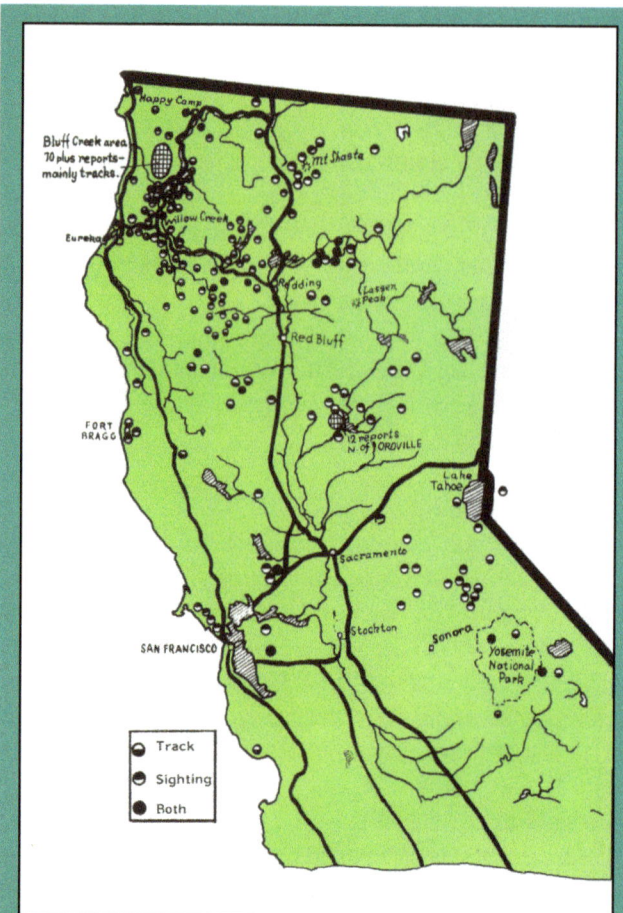

This map of California created by John Green shows reported bigfoot incidents up to about 1980. The Bluff Creak area (crosshatched) was, and continues to be, a prime bigfoot area. California has the greatest number of reported incidents in the United States. It is second only to British Columbia in North America.

OCTOBER 1967

Sun	Mon	Tue	Wed	Thu	Fri	Sat
1	2	3	4	5	6	7
8	9	10	11	12	13	14
15	16	17	18	19	20	21
22	23	24	25	26	27	28
29	30	31				

Holidays and observances: 9: Columbus Day

This type of film required development under the Kodak K-12 process. Seattle, Washington was the closest city to Yakima that had a facility for processing this type of film. We don't know how many film rolls Patterson had with him, but we are certain he had enough rolls to support making a film documentary. Nevertheless, while we can say he had plenty of film, we must remember that the camera held only one roll—about four minutes of filming time. In this connection, his general filming (use of a limited film resource in the camera) does not support a planned hoax involving Patterson. In other words, it is unlikely he would have taken so much general footage (76 feet of himself, Gimlin, the horses, scenery) prior to filming the "creature." As it happened, he hand only 24 feet of film left in the camera when the creature was spotted. Certainly, if he "expected" to see a fabricated bigfoot, he would have ensured that he had plenty of unexposed film in the camera.

10. Timing Aspects

Just exactly when the men departed for California is not clear. Gimlin has stated that the pair were in California for a total of about three weeks. This statement implies that they would have arrived in California on or about October 1, 1967.

However, the article in *The Times-Standard* written on October 21, 1967 from an interview with Patterson states, "Last Saturday they arrived to look for tracks themselves." The date for the Saturday referenced is October 14, 1967. Also, in a Northwest Research Association bulletin issued by Patterson, he states, "Bob Gimlin, a friend of mine, and myself had been in the area a little over a week, when riding horseback up a creek bottom the afternoon of October 20, 1967, we encountered this creature."

I cannot rationalize the timing difference other than to say that men were responding to a footprint report received on September 5, 1967, and it is not likely they delayed their expedition until mid October. Then again, perhaps the "area" Patterson refers to is the specific Bluff Creek area where the creature was sighted, not California per se. Whatever the case, I think we can conclude that if the men were involved in a fabrication, they would have "got it together" on this point to ensure there were no questions.

11. Observance by Others

This information was provided by Peter Byrne. I have no reason to doubt that it is not correct. Given it is correct, the fact that Patterson and Gimlin were seen in the Bluff Creek area on two different occasions by Mr. and Mrs. Hooker has some significance. There is no mention of a third member of the team— a very tall person to play the creature in a hoax scheme. Both Patterson and Gimlin were far too short for this role (creature height is discussed later). One can contend, of course, that such a person joined the two men later.

The movie camera seen here is the same as the one Patterson had. It is about 10.25 inches (27 cm) long, and weighs about 6 pounds (2.7 kg) To change a film roll, one had to be in the dark. Patterson got under a tarp. We can see, therefore, that it was not a quick process—all the more reason to ensure there was lots of unexposed film left on a roll.

This is strip of 16mm movie camera film (it is 16mm wide). There are about 3 and one-half frames per inch, so a 100-foot roll of film would therefore have about 4200 frames. We believe the film was taken at 16 frames per second, and this being the case, a roll would last about 4 minutes, 23 seconds. For a one-half-hour documentary, one would need about seven 100-foot film rolls.

4. The Encounter

OPENING COMMENTS: *Patterson and Gimlin's unusual experience on October 20, 1967, and their activities over the next two days, continue to raise many questions. To our knowledge, the men kept no diaries or journals, nor did they retain any documents resulting from their activities. All we really have is what these men have told us, together with bits and pieces of information that have surfaced over the years. Unfortunately, the process of serial reproduction (one person telling another person and so on) has clouded the facts. There is also the problem that writers and others tend to "make sense" of certain information by more or less "filling in the blanks." It is likely I have fallen prey to some of these conditions, and if so, I will stand corrected.*

On Thursday evening, October 19, 1967, the men set up their camp close to Bluff Creek itself. Gimlin arose early the next morning and rode out of the campsite while Patterson slept in. Gimlin arrived back at the camp at about 10:00 a.m. Patterson was not at the camp at this time. He returned after a little while, and asked Gimlin what area he had covered on his early ride. Gimlin told him where he had been, after which Patterson suggested they re-explore an area they had previously explored. Gimlin agreed, and the men left at about 12:30 p.m.[1]

At about 1:30 p.m. that day, Friday, October 20, 1967, Patterson and Gimlin spotted a female[2] bigfoot down on a Bluff Creek gravel sandbar on the opposite side of the creek.[3] Patterson spotted the creature first, and he yelled, "Bob, lookit!" Patterson estimated the creature to be about six and one-half to seven feet tall, maybe taller, and to weigh about 350–400 pounds.[4] The creature looked in the direction of the two men, no doubt reacting to the thud of hoofs and the clatter of men on horseback.

Patterson's horse reared in alarm at the sight of the creature, bringing both horse and rider to the ground, Patterson pinned below. Gimlin's horse and the packhorse, being led by Gimlin, also reacted. The packhorse panicked and Gimlin released its lead in order to control the horse he was riding.[5] Patterson, being an experienced horseman, quickly disengaged himself,[6] and grabbed his camera. As all of this was happening, the creature walked away, heading in an upstream direction.[7] Patterson ran towards the creature, crossed the creek and took a course parallel to the creature with between 80 to 100 feet or so between himself and the creature. He stopped at a point determined to be about

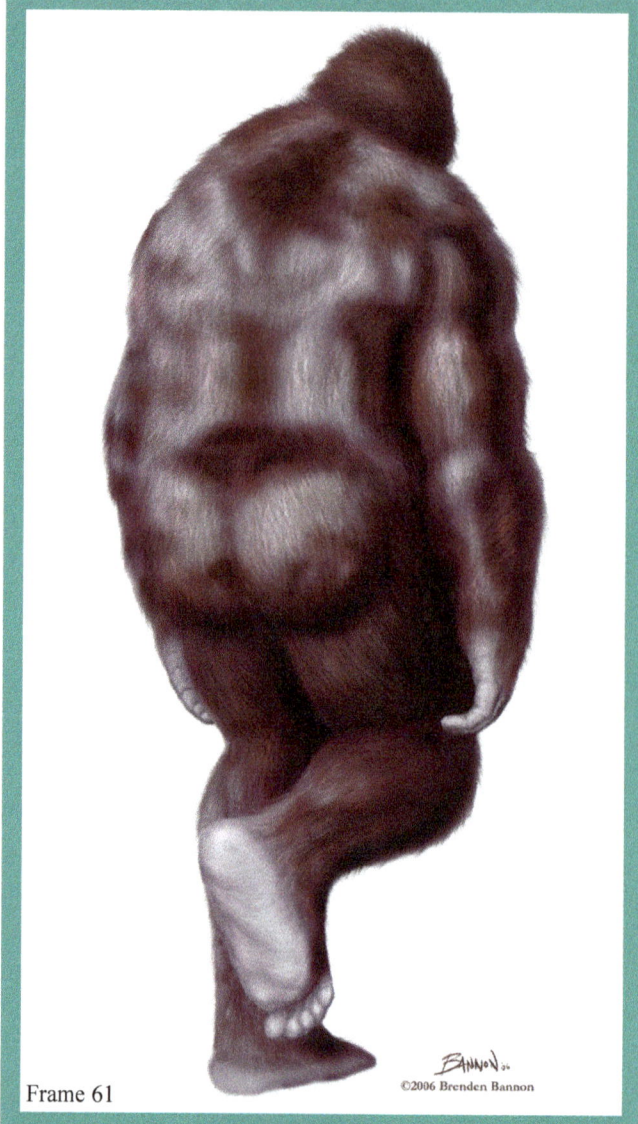

Frame 61

We do not have film showing the creature when it was first spotted. By the time Patterson got himself organized, the bigfoot had travelled a fair distance. Nevertheless, the first frames of the film show the creature from the back. Patterson managed to film it clearly just before it disappeared in the forest fringe. Shown here is the first of six images created by Brenden Bannon, a professional artist. I asked Brenden to study the film frames and create images as though the creature were about ten feet away. Also, to artistically complete an image if part of it were obliterated by forest debris. Brenden studied the film for many months, and I firmly believe his creations depict very accurately how the creature appeared, without the fuzziness of the actual film frames. What we see here, then, is how the creature appeared from the back as it walked away. The film frame depicted is frame 61, as indicated in the lower left corner.

102 feet from the creature and yelled at Gimlin, "Cover me!" He filmed the creature as it passed in front of him, and as it moved further away to the right, he moved closer by about 36 feet (he moved up to the large log we see in the foreground of the film frames). While running, stumbling, stationary, and later walking, Patterson took 24 feet of film footage, which exhausted the film roll in the camera.[8]

In the meantime, Gimlin, on horseback, at Patterson's command took his rifle in hand, and rode slowly towards the creature. Gimlin crossed the creek and dismounted. He then observed the whole scene, rifle in hand, in case his friend was attacked by the creature. The men had previously agreed that under no circumstances would they shoot a bigfoot unless to protect themselves or each other.[9]

Early in its passage, the creature again looked at the men (or possibly just Gimlin), although this action was not caught on the film. Directly in the creature's path there was a slightly twisted wood fragment. The creature either stepped on, or nearly stepped on this fragment.[10] A few steps after this point, the creature turned, and again looked at the men.[11] The film shows this action and we can see that the creature appears to be highly concerned with the men's presence. The creature then hastened its pace somewhat as it continued its passage into a sparsely wooded area.

Patterson continued filming the creature as it disappeared and reappeared between trees that are both in the foreground (close to the camera) and in the distance. Patterson ran out of film with the creature still marginally visible.[12]

After the creature was out of sight, Gimlin wanted to immediately continue pursuit on horseback and proceeded to do so. Patterson, however, did not have his horse or his rifle, and did not want to be left alone. He therefore yelled at Gimlin to return, which he did.[13]

After Patterson's situation was rectified, and he had reloaded his camera, the men then followed the path taken by the creature. They found scuff marks in the gravel and in the creek bed, which may have indicated the creature ran when it was out of sight. They continued up the creek and observed a rock with a wet half footprint on the surface. From that point the path led up into the mountains.[14] The men then returned to the film site and examined the path the creature had taken along the sandbar. Footprints made by the creature were highly evident. In that part of Bluff Creek, there is a sandy clay soil with a blue-gray tinge. This type of soil holds footprints remarkably well for a long period of time.

Frame 307

As the bigfoot emerged out of the forest fringe, it appeared as we see in this image. Considerable musculature is noted, which is one of the main points that preclude the notion that the creature was a man in some sort of costume.

The several clear film frames in this sequence show conclusively that the creature had a flexible foot. Work performed by Marlon Davis shows the foot bending in a natural way, and we can even see sand kicked up by the feet as the creature moves. Of particular note are its splayed toes, which make its calf muscles bulge prior to weight transfer. Many sasquatch casts show the toes greatly splayed, and this was originally looked upon with some suspicion. René Dahinden was quite adamant that such casts did not show that a natural foot was involved. Dr. Grover Krantz, however, was just as insistent that the casts were from authentic sasquatch prints.

Both men then immediately went back to their truck, a distance of fewer than two miles, to get plaster to make casts of the footprints. Upon return to the film site, Patterson made plaster casts of the left and right feet.[15] While the plaster was in the prints, he took movie footage of the tracks in a series. The footprints measured 14.5 inches long by 6 inches wide and were impressed in the soil up to one inch. Gimlin jumped off a stump to see how far his footprints would sink into the soil in comparison with the creature's prints. The results were that the creature's footprints were deeper. Patterson took movie footage of this experiment and the cast-making activities (the men filmed each other). The second film roll, therefore, contains both the creature's footprints in a series and this later footage.[16]

Frame 323

DETAILS AND ANALYSIS

1. Early Morning Activities and Decision

The information given here was provided by Bob Gimlin at a videotaped interview conducted by John Green on March 29, 1992. The fact that Patterson was not at the campsite when Gimlin returned, together with Patterson's suggestion to re-explore a certain area, *might* be considered *curious*. As it happened, the men did see a bigfoot in this area. If Patterson alone had been part of a planned hoax, however, it is unlikely he would have double-tracked the scene. Just how far the men were at this time from the sighting location appears to be about one half hour by horse. We can gather this from the fact that the men had lunch at their campsite and then set out for the designated area, spotting the creature at about 1:30 p.m. Certainly the location was too far away for Patterson to have gone there and back that morning to ensure something was "set-up," as it were. This conclusion, of course, does not eliminate the possibility that Patterson met someone close by who then went directly to the film site. However, by far the best explanation for Patterson's absence is that he was simply down by the creek washing or getting a drink, as Gimlin stated in his interview with John Green, and has confirmed with Thomas Steenburg on more than one occasion.

2. Significance of the Creature being Female

The fact that the creature is female greatly increases the film's credibility. As a rule, monsters of any nature are usually associated with males—a universal theme. This point would have been even more applicable in 1967. Even today, most people are surprised when they are told the creature filmed was female. Furthermore, even when people are looking at a photograph of the creature in the film, its very evident gender does not register because they have been preconditioned to think "male."

The image seen in this frame appears to show that the creature had a skirting of longer/thicker hair about its buttocks. There is some conjecture that this might trap some body waste and might be partially responsible for the obnoxious odor very often sensed when bigfoot are present. We can reason that a creature of this nature would definitely have a strong body odor. I have even entertained the thought that it might bath in hot springs and thereby retain a "rotten egg" smell (which truly takes one aback) caused by the the sulphur in the water. Indeed, people have described the creature's smell as exactly like rotten eggs. Some monkeys in Asia, incidentally, are known to bathe in hot springs.

I will also mention that gorillas in Africa do have an offensive odor, which actually increases when they are agitated. In other words, it appears to be a bit of a defense mechanism.

What I am saying here is that it is unlikely someone would have thought to make the creature a female in a hoax scheme. Certainly, if the creature were a fabrication, making it female was a stroke of genius. Nevertheless, there is a minor side issue here. Making the creature female ment that one could "get away" with a shorter/smaller person to play the part (one within the human range) as females of most primate species are usually shorter and smaller than males.

3. The First Observation of the Creature

Some documentation states that the creature was squatting beside the creek when it was first seen by Patterson and Gimlin. The only firm reference I can find in this regard is in Patterson's statement to a reporter for *The Times-Standard*, Eureka, California. Patterson is quoted as saying, "I yelled 'Bob Lookit' and there about 80 or 90 feet in front of us this giant humanoid creature stood up." The key here, of course, is in the words "stood up." Naturally, if the creature stood up, it had to be bending over, crouching, kneeling, squatting, or sitting prior to standing. In another "first observation" statement, Patterson is quoted as saying the creature **stood** there for a short time and then crossed the creek. Gimlin stated most definitely that the creature was "getting up" when he and Patterson first saw it. I think we can reasonably say that it was in the process of standing from a squatting position when the men first saw it.

On a different note, the fact that the creature was beside the creek is important. We can conclude from this information that the creature probably had wet feet that picked up sand and soil as it walked on the sandbar. The soil in the Bluff Creek area has a light blue-gray tinge. It has been reasoned that these conditions might account for the very light color of the soles of the creature's feet seen in the film. Nevertheless, an experiment conducted by Daniel Perez at the film site indicated that the soil does not adhere to wet human feet for more than a few paces. All I can offer here is that the creature probably had light colored skin on the soles of its feet.

4. The Estimated Height and Weight of the Creature

The creature's actual height and weight have been debated for many years. It is important to know this information because it helps to either support or refute the authenticity of the creature. The taller and heavier the creature is, the less likelihood it was a "man in a gorilla suit." The lowest estimates are a walking height of about 5 feet, 8 inches to 6 feet, 3 inches, with a weight of about 190 to 240 pounds (Dr. E. Sarmiento). The highest estimates are a walking height of 7 feet, 3.5 inches, with a weight of 1,950 pounds (J. Glickman,

Frame 343

This frame and another less than a second before, provide the best insights into the creature's body profile. There are different interpretations as to how the creature actually looked. Unfortunately, the clarity is just not there to get a proper "fix" on its facial features.

The Weight Dilemma

Sasquatch footprints are usually deeply impressed into the ground, which naturally suggests the creatures are very heavy. Indeed, they are perhaps much heavier than we think. As their feet are very large, they would distribute the weight and thereby reduce penetration into the soil (sort of like snow shoes). Naturally, the softer the soil, the greater the penetration. This is a relative factor. Remarkably, prints are often found in soil where prints of a man of fair weight will hardly register. Given the size of sasquatch feet, and the soil compaction, the creature's weight seems to defy logic.

North American Science Institute). I will mention here that the Glickman weight estimate calculation has been disputed by Dr. Henner Fahrenbach, who states that the final figure should be 542 pounds (based on its chest circumference).

Other professionals/researchers have estimated the walking height as follows:

Dr. Grover Krantz	6 feet (about)
Yvon Leclerc	6 feet, 3.5 inches
Dr. Donald Grieve	6 feet, 5 inches
D. Bayanov/I. Bourtsev	6 feet, 6 inches
John Green	6 feet, 8 inches

Dr. Krantz tells us that walking height is between 8% and 8.5% shorter than standing height (when one is standing fully erect—legs straight, back straight, and head straight). We can therefore conclude that, using the lowest percentage (8%), the absolute lowest standing height for the creature in the film is 6 feet, 1.5^1 inches. The lowest weight, as stated, would be 190 pounds. The creature's minimum height alone definitely rules out either Patterson or Gimlin as being the "creature" in a fabrication. Both men were well below 6 feet in height (Gimlin 5 feet, 7 inches; Patterson, 5 feet, 3 inches).

5. The Reaction of the Horses

We are told by both Patterson and Gimlin that their horses and the packhorse were alarmed when the bigfoot was sighted. In all likelihood, it was probably the creature's odor that startled the horses. We must remember that horses are very sensitive to odors and would have picked up the creature's scent very quickly at a considerable distance. It is noted that Patterson says he sensed an unusual odor, however, this fact is immaterial (we don't need to depend on it) because the horses would have detected the creature's odor even if Patterson or Gimlin had not sensed anything. It is unlikely the horses would have reacted to a fabricated creature (i.e., a man in a costume) as there would not have been an odor. Speculating on some kind of offensive artificial odor is going beyond reason. Nevertheless this detail could be a part of a fabricated story.

6. Patterson's Fall

According to Bob Gimlin, when he was interviewed by John Green in 1992, Patterson's horse did not fall, pinning Patterson below. Yet Patterson is quoted in a newspaper article *(Times-Standard)* as saying, "My horse reared and fell, completely flattening a stirrup with my foot caught in it,"(he later showed the bent stirrup to Al Hodgson and Syl McCoy at the Lower Trinity Ranger

Frame 352

The epic moment in the film is when the bigfoot half turns and looks at the men. The scene as we believe it to have been is indicated by the film-site model. Patterson was standing in the position indicated by the red pin. The bigfoot gave the men a somewhat concerned and ominous stare. It did not want anything to do with them, so simply kept moving.

The detailed drawing on which the model is based is at the following link:

http://forum.hancockhouse.com/article.php/20050924140836229

---

1. This is the absolute lowest estimate (5 feet, 8 inches) multiplied by 1.08 (i.e., addition of 8%).

Station). I have no way to reconcile this discrepancy other than to say that Gimlin simply did not recall or did not see the fall. We can reason, however, that it is highly unlikely Gimlin would have contradicted Patterson's account if both men were involved in a conspiracy. In other words, why say anything that would lead to questions on the incident?

On the other hand, if only Patterson was involved in a fabrication, why on earth would he lie about falling when Gimlin was right behind him? What I have concluded is that Gimlin was taking care of the reaction of his own horse and the packhorse when Patterson's horse fell. In other words, Gimlin was not looking at Patterson at that precise moment.

Whatever the case, we will still examine the "fall." A regular horse falling on a man would probably result in serious injury to the man. However, in Patterson's case, he was riding a small quarter horse (not his Welsh pony, "Peanuts," which he normally rode at home—just a small regular horse). The weight of Patterson's horse was less than 800 pounds. Patterson himself was not a large man. He was about 5 feet, 3 inches in height with a slight build. As such, he was able to use small horses. It is conceivable, therefore, that Patterson escaped serious injury because of the small size of his horse.

In any event, the incident has only marginal significance in the story. It merely reinforces the point that Patterson's horse was very alarmed with the presence of the bigfoot.

7. The Creature's Awareness

The questions have been asked as to (1) why the creature did not sense (smell or hear) the intruders and disappear before it was seen, and (2) given it did not sense the men, why did it not dart off into the nearest forest cover as soon as it saw them?

On the first question, normally, animals can hear and smell other animals from a great distance, as previously discussed. In this case, however, the bigfoot was near a rippling creek, which probably muffled the sound of the approaching horses. Also, the men had approached the creature from the upstream direction (opposite to the stream flow). As air currents move in the direction of a stream, the odor from the men and their horses was minimal. We can also speculate a little that the bigfoot was thoroughly preoccupied, *possibly* looking for fish or other creatures in or near the creek. It may be reasoned, therefore, that when the bigfoot was seen, it probably did not *immediately* know it was being observed.

Frame 364

It is in this frame that I detect the high concern the creature has with the men's presence. Although the time between this frame and frame 352 is less than one second, in my opinion, the creature quickly goes from a state of full awareness that it is being observed to one of apprehension. Given I am correct, one thing that surprises me a little is that after this point, the creature does not look back to see if it is being followed. I will mention here that in the William Roe incident (1955) the creature glanced back as it walked away, and I have read other sighting where this has occurred.

On the second question, when the horses reacted (making excessive noise), probably causing the creature to look in the men's direction, it may have just seen horses, not men on horses. Even Patterson's yell to Gimlin (if heard) may not have alerted it to the presence of men. Horses were not a threat to the creature. It had probably experienced such a reaction from horses before, resulting in the horse or horses simply running away.

Then again, it is entirely possible that the creature was fully aware that the men were coming in its direction but did not move until they got beyond its "comfort zone" (i.e., the men got too close.) Most animals have such "comfort zones," which vary based on the creature's speed and agility.

Certainly, it is reasonable to conclude that a real bigfoot would have immediately run into the forest. However, it must be kept in mind that the normal reaction in hundreds of bigfoot sightings is that the creature simply walks away when it realizes it is being observed by people, and this is exactly what the creature did at Bluff Creek.

Nevertheless, while we can justify the creature's lack of awareness or slow reaction, one might conclude that such was a necessary part of a fabrication to provide an opportunity for filming.

8. The Filming

Patterson ran out of film while the creature was still marginally visible in the distance. The segment of film (about 24 feet) showing the creature comprises 953 frames and is what we refer to as the Patterson/Gimlin film. As previously pointed out, the first 76 feet of this film roll was used for general filming. I stress again that it does not appear Patterson was overly concerned with the amount of unexposed film left in the camera at any particular time. If the sighting had been "planned" with Patterson's knowledge, he would likely have ensured the camera had plenty of usable film when the incident occurred (although not necessarily).

9. The Pact Not to Shoot a Bigfoot

Some people interpret this pact as a ploy on Patterson's part. In other words, they contend that Patterson alone was involved in a fabrication. The pact was to ensure the safety of the "man in the costume" if Gimlin overreacted upon seeing the "creature." We can reason, however, that this is a very weak argument. There could be no guarantee that Gimlin would not shoot the "creature." He was an experienced hunter with a powerful rifle. No person in his or her right mind would agree to be the "creature" under these conditions. The fact that

This photograph shows the soil at the film site. It records footprints much like sea sand, however it contains particles of gravel and loam that give it more body. It might be reasoned that such allows it to hold impressions even after severe rain storms. The area is not totally clear as we see here. There are rocky sections and a lot of forest debris. Nevertheless, I am sure there were enough clear spots for many good footprints to register.

The bigfoot was most definitely female from what I can see, but at least one anthropologist has difficulty with this, and another says it is a male despite evidence to the contrary. Its body proportions, however, appear to be very different from a human female as we can see in this comparison. If I were to guess as to the age of the creature, I would say about 35. I would also guess that we are seeing her at about her greatest weight. The film was taken in late October, so we are seeing her after a summer of plenty.

Further insights on this comparison are at the following link:

http://forum.hancockhouse.com/article.php/20060611103140663

Gimlin has firmly stated he definitely could have shot the creature reinforces this point.

Nevertheless, the men left themselves in a difficult position by having rifles at their disposal. At that time, *Life* magazine had offered a $100,000 reward for a bigfoot, dead or alive. It is seen, therefore, that the rifles lessened the credibility or authenticity of the encounter, (i.e., if it were a real bigfoot, why not shoot it and get the reward?). On the other hand, the rifles increase credibility. If the men were part of a fabrication, why would they leave themselves open to questions on not shooting the "creature" for the reward by having firearms available?

There is also another reason for the "don't shoot" pact. Patterson expressed fear of repercussions from First Nations people. Many of these people hold a strong belief in bigfoot creatures and give them certain spiritual attributes. There would definitely be a reaction by some First Nations people if one of the creatures were killed. Patterson also mentioned that he was concerned about the reaction of other bigfoot that might be in the area if one of their number were shot.

In any event, the "pact" was essentially useless. There is no way it could guarantee that either men would not shoot one of the creatures if he saw fit for any reason, or no reason. Indeed not long before Patterson died (and after years of much stress brought about by his experience), he told Peter Byrne: "Peter, we should have shot that thing, then people would have believed us."

10. The Wood Fragment

When René Dahinden visited the film site in 1971, he carefully compared the film frames to the actual film site. He observed that what appeared to be the same wood fragment seen in the film frames was still in place. Dahinden used the fragment location to determine the distance of the creature from the camera. He arrived at 102.8 feet, a figure later reasonably confirmed (102 feet) by Dr. Grover Krantz.

Upon leaving the film site, Dahinden picked up the wood fragment and took it home with him to Richmond, British Columbia, Canada. The fragment is about 26.25 inches long, by 3 inches wide, and 1.5 inches thick (width and thickness vary).

René showed me the fragment in 1995, and I believed it could be used to perform a calculation to approximate the creature's height. At that time, I was not aware that nearly four years had elapsed between the filming and the time Dahinden obtained the wood fragment. I was definitely taken aback when I learned of the time lapse. That a piece of wood this size could

The Killing Question

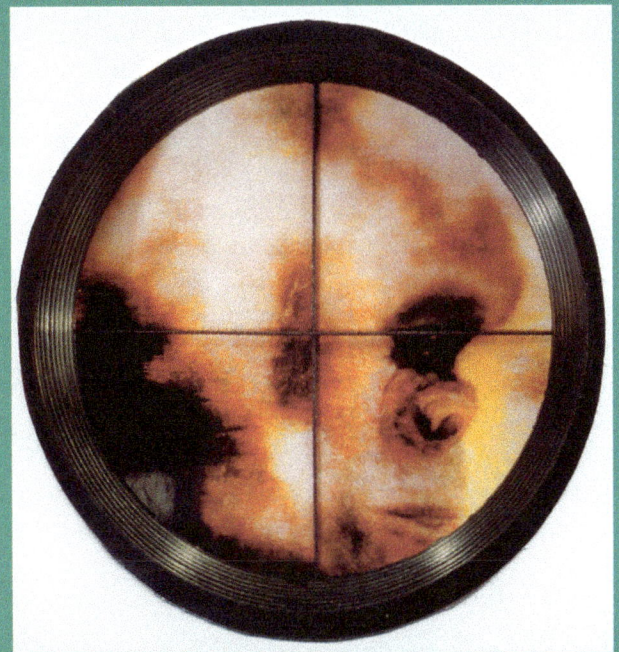

The issue of whether or not one should shoot a bigfoot has been debated to the point of exhaustion. Most certainly the creature has been shot at and hit many times over the past 200 years. Sometimes it has been brought down, but unfortunately the body has not been brought in for scientific analysis.

The image I show here was created by Yvon Leclerc. It is based on frame 339. I added a mocked-up rifle scope to "get the feel" of what a hunter would see when he leveled his rifle on the creature. Here I will point out that the cross-hairs are on what I believe is the exact spot a bullet needs to penetrate in order to kill the creature. Shots to the body, front and back, have not brought it down. However, I am going to state most emphatically that if I saw in my rifle scope what is seen here, I would not be able to pull the trigger. Perhaps one day rifle scopes will be equipped with a digital camera, giving the hunter the option to shoot a photograph instead of a bullet, and we will be able to see the one that "got away."

remain in exactly the same spot for nearly four years is highly questionable. **Nevertheless, it is not impossible**.

I recently performed a comparison of a film site photograph taken by Peter Byrne in 1972 with an actual film frame photograph. Many wood fragments, fallen tree branches and other debris were still in the exact same place—some much smaller than the fragment retrieved by Dahinden. I was also able to determine that the position of the fragment in a 1971 photograph taken by Dahinden was very close to (if not exactly the same as) the position of the fragment in the actual film frames. Finally, a direct photographic comparison of the fragments shows they are probably one and the same. I am now quite certain the fragment Dahinden retrieved is the same fragment.

The fragment is very close to the creature in frame 307 of the film and is seemingly lying very straight (i.e., parallel to the creature's path). I am told, however, that without knowing the exact camera angle and fragment angle, creature height calculations based on the fragment are meaningless. One authority has told me, however, that this is not the case and that a reasonable calculation could be performed.

11. The Creature's Turn Towards the Men

The way in which the creature turns is highly noteworthy. As it does not have a neck, per se, it must turn its entire body to see anything at right angles (i.e., directly to its right or left). This anomaly does not rule out a hoax, but it is another factor that gives some credibility to the creature's reality. It should be noted, however, that recent film analysis does show partial head turning.

12. The Passage and Pace of the Creature

Although the creature did walk away after it was first spotted, it did not disappear into the forest. It headed towards the forest and then veered right and more or less walked along the forest fringe. After deliberately looking at the men, or just Gimlin (second look) during its route, it still continued forward at a steady pace.

During all of this time and up to the time the creature turned to look at the men a third time (counting its initial look), it is very evident the creature was not overly concerned with the men's presence. It was simply just concerned enough to walk away. However, when it turned and looked toward the men (last look), its facial expression seems to me to show high concern and it hastened forward. Now we have a totally different situation—something had evidently heightened the creature's concern and it wanted to get away from the men. We can speculate that it became more concerned upon hearing Patterson's camera and then seeing the camera pointed at it. We might also note that at this point the

The Wood Fragment

Seen here are several shots of the wood fragment retrieved by Dahinden. I have measured, photographed, displayed, and pondered this artifact for many years. I have stated that I believe it could be used to determine the height of the bigfoot seen in the film. The calculations I have performed support the findings of Jeff Glickman that the creature was over seven feet tall. However, the camera angle and fragment angle need to be considered, and I have not compensated for these factors. Nevertheless, in my opinion, the adjustment to my calculation would be minimal.

creature saw the men on foot and came to the realization that it was being pursued by men—previously it saw horses, now it saw men.

Whatever the case, many film skeptics point out that at this moment in particular a natural creature would have run to the nearest forest cover. The film frames *indicate* the forest to the creature's left was the nearest cover. We get the false impression that this forest area is only a few feet away from the creature, but in reality it was at least 88 feet away. Also, while it is reasonably level back there for some distance, the level ground is terminated by a mountainside. Nevertheless, cover was available in this direction. However, there was also cover directly ahead of the creature, and **most importantly**, after a few more steps, a little "island" of trees and mangled underbrush to the creature's right came between it and the intruders. This little "island" would definitely hamper a clear rifle shot, so the creature's decision to continue forward was probably its best option.

But why did the creature just hasten its pace rather than run? The worst thing a human can do is immediately run if suddenly confronted with a wild animal such as a bear. This is something we have learned from the animals themselves. Such action sparks an immediate reaction on the part of the aggressor that can prove fatal. Perhaps therein lies the answer. Given this analysis, it appears evident the creature was unaware of firearms. Guns turn the slow retreat tactic into an advantage for the hunter.

On the other hand, it can be reasoned that the creature's passage and pace indicate a hoax. One can say that the "creature" was displaying itself for the purpose of the filming. On this point we might again think about the chance taken by the "creature" if it was a fabrication. There was no guarantee that Patterson and Gimlin were the only men in the area (i.e., there may have been hunters around).

After the creature disappeared from view, later research by Bob Titmus indicates that it turned right and crossed the creek. It went straight ahead for a short distance and ascended a hill where it possibly sat down for a few moments and observed the men, who were still down on the creek bank (Titmus observed a flattened-out spot where he believes the creature sat down). It then continued upstream. Bob Titmus was a taxidermist and veteran bigfoot investigator. He went to the film site on October 29, 1967. Titmus was able to follow the creature's footprints, which were still very evident even though nine days had passed since the sighting.

13. The Decision not to Immediately Pursue the Creature into the Forest
Patterson's decision to stop Gimlin's immediate pursuit of the creature can certainly be justified. The men had been told of three different sizes of footprints in the area, so the creature they filmed might not have been alone. Also, as the creature they filmed was female, the likelihood of

Shown here is the film-site model from three different angles: Top: from left; Center: from right; Lower: from back. In the first and third image, Patterson's position is indicated by a blue dot. In the first two images, the creature is shown by a little red box (angle of the little cut-out makes it nearly invisible in these photos). The images give us a bit of a feel as to the action that took place. The creature did not travel directly ahead. It veered to its left toward the leaning tree seen clearly in the lower image.

a male being nearby increased their concern. Without his horse and rifle, Patterson would have been left in a very dangerous situation. Both men were impressed with the size of the creature filmed. Upon seeing that the creature was female, Gimlin remarked to himself as to the size of her "husband." Aggressively pursuing the female would definitely provoke the male. Finally, the fact that Patterson probably had photographic evidence of the creature highly supports his decision. In other words, why chance a confrontation if you have sufficient evidence?

Nevertheless, we must also consider hoax factors. It might be contended that Patterson did not want Gimlin to pursue the creature because he (Patterson) knew the creature was a fabrication. This contention is possible, I suppose, but it does not seem probable.

14. Following the Footprints
Neither Patterson nor Gimlin mentioned the little hill in their accounts, nor did they mention the spot where the creature possibly sat down. In all likelihood, they did not consider the hill important (they did not have a good look around) and simply missed seeing the sitting spot.

15. The Plaster Casts
The fact that Patterson and Gimlin made footprint casts indicates the men believed the creature was real. It also indicates that if the creature were a fabrication, these men were not involved. Patterson probably had the creature on film. If the men were involved in a hoax, would they have really bothered making casts? It is obvious they made the casts for two reasons: first, to substantiate the film; second to at least have tangible evidence of the sighting if none of the film they took turned out. Nevertheless, if they had not made footprint casts, they would have been questioned on this point.

However, they could have come up with some excuse for not making casts. Keep in mind that, to my knowledge, there are no photographs of footprints or casts directly associated with other alleged bigfoot films or videos (Ivan Marx film, Freeman film, Redwoods video, Snow Walker video, Memorial Day video). Furthermore, there are no footprint photographs or casts that are directly associated with any hoaxed still photographs of the creature (or *any* bigfoot photographs for that matter). I believe bigfoot hoaxers do not provide such additional evidence because it is difficult to fabricate footprints without possible hoax indicators. I discuss this issue further in relation to the second film roll (Point 16, following).

16. The Second Film Roll
In any hoax scheme, it makes sense to *minimize* the amount of information or evidence one provides on the event. The more one provides, then the greater the chance of evidence being uncovered that points to a hoax. Patterson and Gimlin went totally overboard in this connection. Not only did they film the creature's footprints and make casts, but they also filmed their cast-making activities. We can only conclude here that if the creature were a hoax, then Patterson and Gimlin were not aware and really thought they had seen and filmed a natural bigfoot.

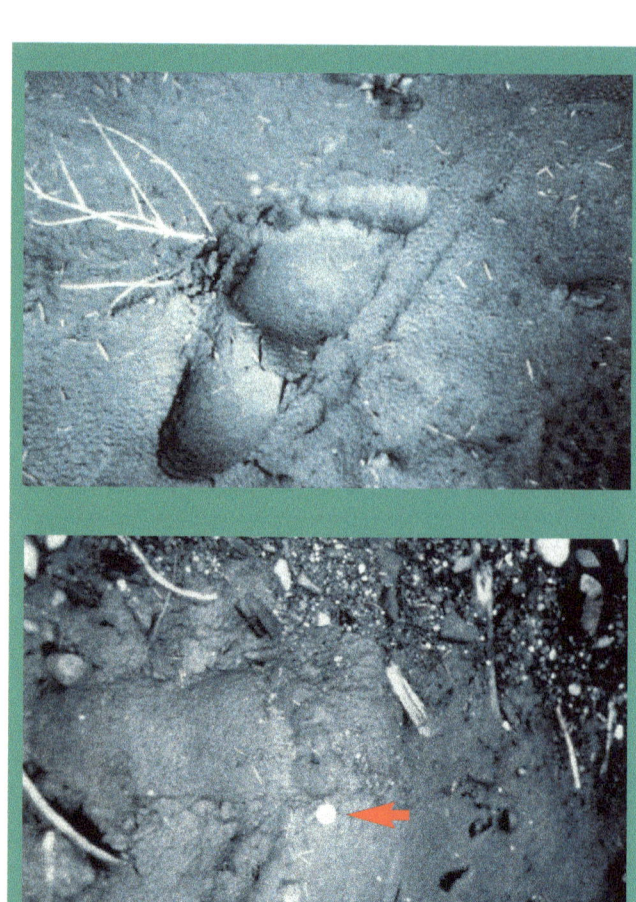

Shown here are photographs taken by Robert Lyle Laverty, a timber management assistant, on October 23, 1967 (three days after the filming) of the footprints left by the creature at the film site. The prints measured about 14.5 inches long and were impressed in the soil up to a depth of about 1 inch. The center photo shows an American twenty-five-cent coin adjacent to the big toe to indicate size (red arrow).

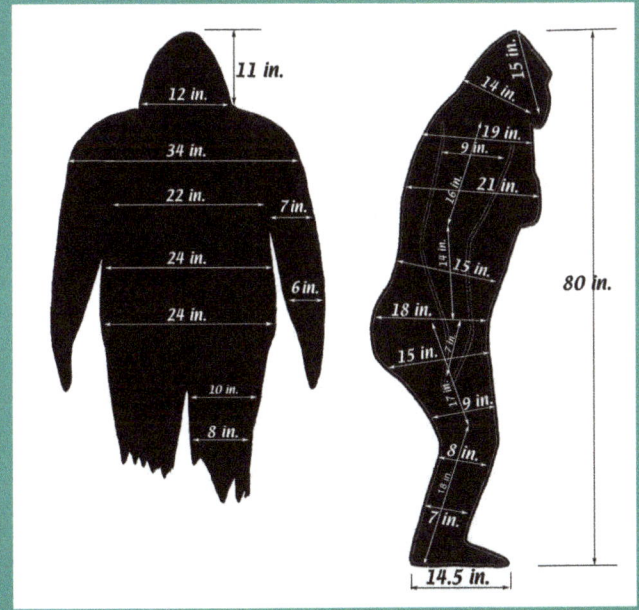

Seen here are the creature's height and body dimensions as estimated by John Green. John based his calculations on the creature's foot size of 14.5 inches as indicated by its footprints. The height was verified by photographing a person at the film site from the same distance as the creature seen in the film, and then registering both images. Since the film image of the creature is blurry and it has hair or fur of unknown thickness, the measurements are not exact. What we see here is the creature's *walking height*. Its *standing height* would be up to 6 or 7 inches taller.

This is full film frame 352. I have shown the distances from the camera to various objects and the creature. The objects are as follows:

Group of Stumps:	174 Feet	Upright Stump:	190 Feet
Single Stump:	170 Feet	Tree One:	85 Feet
Creature:	102 Feet (Krantz)	Tree Two:	140 Feet
Long Branch:	62.8 Feet	Tree Three:	95 Feet
Log:	36 Feet	Leaning Tree:	146 Feet

(Top) Diagram of the film site by Bob Titmus.[1]
(Lower) Model of the film site showing the scene in Frame 352.
(Right) Frame 352 and model showing approximate relationship of the creature and other objects. One will notice that the relationship of the objects in the film frame is highly deceiving.

[1]. Concern has been expressed on the accuracy of this diagram. Nevertheless, save the revision I note, I believe it is reasonably accurate.

The top image shown is full frame 352 (as a reminder, the film was taken in 1967). The lower image is a photograph taken by Peter Byrne in 1972 showing Michael Hodgson with a measuring pole in about the same position as the creature. Given Hodgson's known height, a perfect registration of both images using the width of the dead tree directly in front of Hodgson compared with the same tree in the film frame reveals that the creature's height was 7 feet, 3.5 inches. One might also note that although the images are about 5 years apart, the film site was hardly disturbed save two trees that had fallen down. Many debris objects are in exactly the same place, even those that are quite small. This observation serves to confirm that the wood fragment recovered by René Dahinden in 1971 could definitely have been the same fragment as seen in the film frames.

The Patterson/Gimlin film site, Bluff Creek, Six Rivers National Forest, Northern California (as indicated), from 6372 feet. The site's general location is about 38 miles south of the California/Oregon border, and 18 miles inland from the Pacific Ocean. In 1967, the nearest paved road was about 25 miles away. (Photo: Image from Google Earth, © 2005 TeleAtlas; Digital Globe)

The film site today is totally different from that shown in the film-site model. Bluff Creek is fed by the run-off of many mountains to the north. At one moment the creek is just a babbling brook, and the next moment a raging torrent. I believe that is how the creek got its name. It appears the creek had "one of those days" and tore out about two-thirds of the film site. All that is left is a deep gorge, as seen in the photos on the left, and a small clearing. For the first photo (left), I stood in the gorge. The remains of the site are in the background. The lower photo shows the extent of the gorge, again facing the site. In the photo above, I am standing in a part of what is left of the clearing. I believe that where I am standing is far to the left of what is seen in the model or frame 352 of the film. I took these photos in 2003. Daniel Perez has informed me that the site has drastically changed even more in recent years.

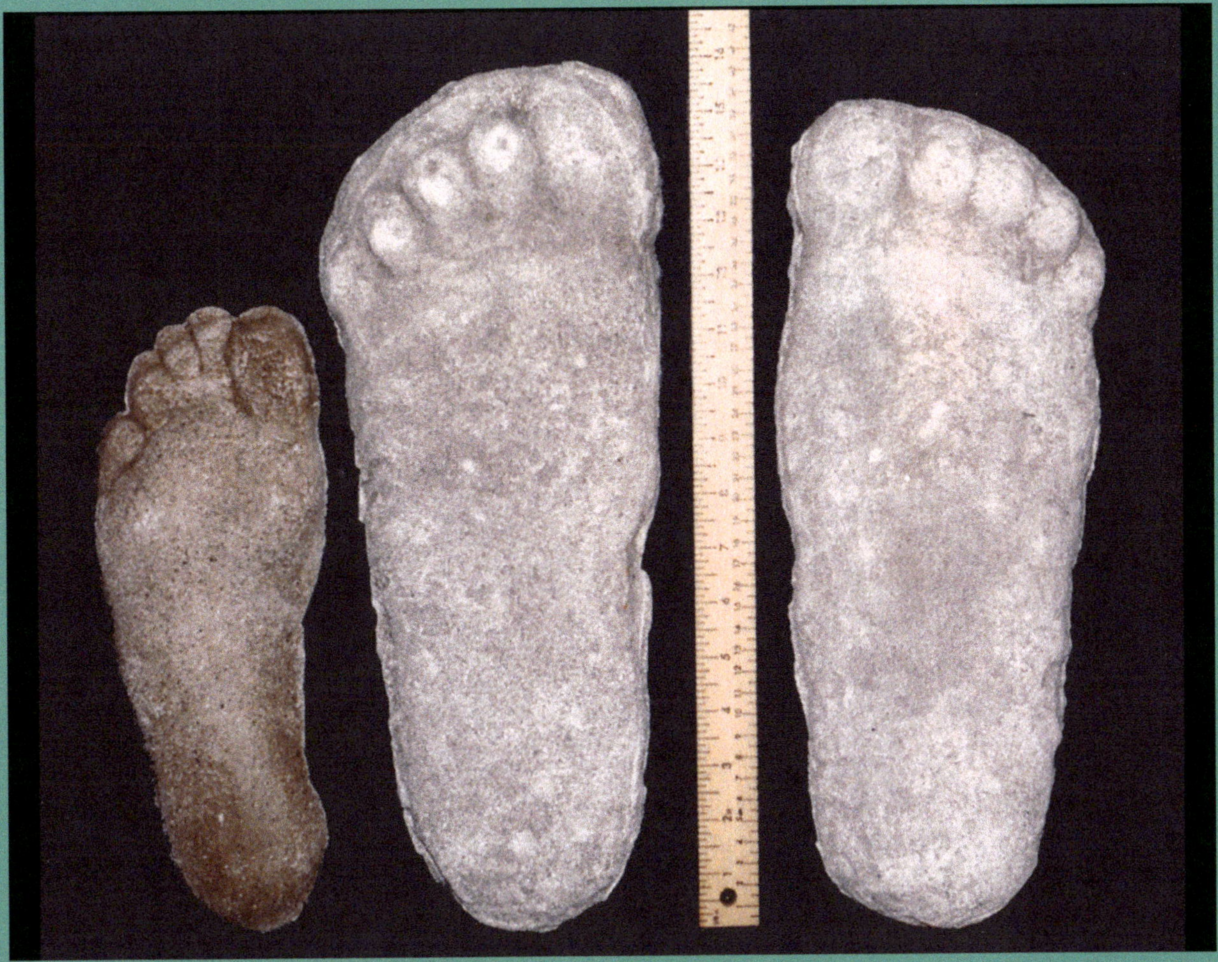

Copies of casts made by Roger Patterson on October 20, 1967 from the tracks the creature left at the film site, with the cast of a human foot on the left for comparison purposes.

Four casts made by Bob Titmus from footprints found at the film site on October 29, 1967. Titmus made a total of ten casts from consecutive prints.

A 3D scan of the extreme right cast may be seen at this link (image provided by Dr. J. Meldrum): http://forum.hancockhouse.com/images/articles/20080425134353577_1_original.jpg

5. Crucial Decisions

OPENING COMMENTS: *At this point, we have Patterson and Gimlin at Bluff Creek contemplating their next move. Patterson has two rolls of exposed film—one roll showing the creature, the other showing activities after the creature left. Considering the fact that they had followed the creature's footprints for up to one quarter mile or so into the wilderness, the earliest time would be about 3:30 p.m. As it will be seen, Patterson and Gimlin made a number of decisions that are difficult to reconcile without detailed analysis and explanation.*

Patterson was eager to get his film of the creature developed **to ensure that he had in fact caught the creature on the film.** On this point, Gimlin has stated, "We weren't sure from Roger stumbling and falling down on the sandbar and getting up and running; we didn't even have an idea that we had anything on the film at that time…in fact it was doubtful that we did have anything."

The men therefore decided to airship the films (both rolls) to Al DeAtley, Patterson's brother-in-law in Yakima, for immediate processing.[1] The plan was to wait for word from DeAtley as to what, if anything, was on the main film. This information would dictate their next move. In other words, if they had <u>not</u> captured the creature on film, they would stay longer and try again.[2] The men decided they would **both**[3] travel to a shipping facility to make the shipment. This task accomplished, they would then return to their campsite.

Leaving their horses tethered at their campsite,[4] the two men started out in their truck for a shipping facility at about 3:30 p.m. They drove directly to Willow Creek, arriving shortly after 6:00 p.m, and went to a store owned by their friend, Al Hodgson. They probably planned to telephone Al DeAtley from this store to inform him of events and discuss shipping the films to him. However, as it was after 6:00 p.m. the store was closed.

Using a pay phone right outside the store, they telephoned Hodgson at his home. Hodgson came directly down to the store and met with the two, whereupon they related their sighting experience to him and then probably called Al DeAtley.[5]

As they were anxious to ship the films, we believe they asked Hodgson to telephone Don Abbott (a cultural anthropologist with the British Columbia Provincial Museum), and ask him to come down to the film site with tracking dogs. Furthermore, as the men met with

Al Hodgson in 2005. (Photo, D. Perez)

Willow Creek in 1960. Small towns don't change very much, so I am sure it appeared much the same when Patterson and Gimlin rolled in on October 20, 1967.

Hodgson again that evening at the Lower Trinity Ranger Station, it appears they planned to do so at this time (meeting discussed below).

Patterson and Gimlin then probably drove to Murray Field, a commercial air shipment faculty just north of Eureka, and shipped the films to DeAtley. Murray Field is the closest air facility to Willow Creek and Bob Gimlin recalls going to an airport.[6]

The Murray Field Airport, just south of Arcata.

During this time, Al Hodgson telephoned Don Abbott and relayed Patterson's request. Abbott, however, declined to come down to the site, stating that he would wait to see the film. After talking with Hodgson, Abbott informed John Green of events and stated that Patterson had asked for tracking dogs. Abbott also telephoned Al DeAtley and requested that the film be brought to Vancouver, B.C., for viewing by scientists at the University of British Columbia. DeAtley promised he would discuss Abbott's request with Patterson.[7]

Inside the Murray Field Airport office.

On their way back to their campsite, Patterson and Gimlin stopped at the Lower Trinity Ranger Station as planned, arriving at about 9:00 p.m. Here they met with Syl McCoy (another friend) and Al Hodgson. Patterson and Gimlin again related their sighting experience.[8]

At about 9:30 p.m. Patterson contacted a reporter for *The Times-Standard* newspaper at Eureka, to whom Patterson related the filming experience in considerable detail.[9] An article (the first published on the incident) appeared in the newspaper the following day, October 21, 1967.[10] The men then immediately returned to their campsite.[11] By the time they arrived, bad weather was closing-in. By about midnight, it was raining heavily.

Type of plane that would have been used for the shipment.

Fearing a possible landslide on the Bluff Creek road, Patterson and Gimlin decided to get out of the area. At about 5:30 a.m. (October 21, 1967), Gimlin went to the film site and covered some of the creature's tracks with bark to protect them. He had brought cardboard boxes from Hodgson's store for this purpose, but had left them out in the rain and they were too soggy to use. The men then packed up and left for Yakima. They experienced great difficulties getting out of the area. The Bluff Creek road had caved away so they had to take the Onion Mountain route.

After getting word from Don Abbott, John Green immediately tried to contact René Dahinden, who was in San Francisco. Dahinden was not at his hotel, so Green left an urgent message for him to contact Al Hodgson at Willow Creek.[12] After talking with Hodgson, Dahinden traveled immediately to Willow Creek, arriving at about noon, October 21, 1967. Here he met with Jim McClarin, another bigfoot investigator, at Hodgson's store. A short time later, Patterson, on his

Jim McClarin is seen here on the left with me at the Willow Creek Bigfoot Symposium in September 2003. We are standing in front of Jim's famous bigfoot carving that he made in the late 1960s. It is now placed in front of the Willow Creek–China Flat Museum, California.

way back home, telephoned the store from Orleans (about 26 miles north of Willow Creek). He talked to Dahinden and informed him of events. Patterson stated that the pair had left the Bluff Creek area in view of the bad weather conditions. He evidently told Dahinden that the film (i.e., film of the creature) had been sent to DeAtley for processing, as Dahinden and McClarin thereupon headed out immediately to Al DeAtley's home in Yakima to see the film.

Al DeAtley received both film rolls in Yakima on Saturday morning, October 21, 1967. He probably drove to Seattle and had the films developed that same day at the Kodak Technicolor Laboratory in that city,[13] and then drove back to Yakima.

Patterson and Gimlin arrived at Yakima late Saturday night, October 21, 1967 or early Sunday morning, October 22, 1967. John Green arrived at DeAtley's home in the early afternoon, Sunday, October 22, 1967, and he and DeAtley awaited the arrival of Patterson. When he arrived, DeAtley took him alone to the basement and showed him the film of the creature, and possibly the second film. The film showing the creature was then shown several times to John Green. Bob Gimlin was not present. He was at home resting, having driven the entire way home.

Dahinden and McClarin arrived at Al DeAtley's home in Yakima at about 3:00 p.m. the same day (Sunday, October 22, 1967). The film of the creature was shown to the new arrivals and to the others present (again several times).[14] The group then discussed how Patterson and Gimlin should go forward with the new evidence. Patterson did not show the group the general movie footage he had taken (i.e., the first 76 feet of the first film roll), nor did he show the other footage on the second film roll.[15]

The film of the creature impressed the researchers in that it appeared to be a natural animal that matched the description of a bigfoot. They were not impressed with the creature itself, as they were expecting something much more spectacular. Seeing that "little dark thing" run across the screen was hardly in accord with their preconceived images of a bigfoot. Nevertheless, nothing appeared to indicate that Patterson was being untruthful about his experience. At some point after this session, Patterson had duplicates made of the main film and put the original away for "safekeeping."[16]

Syl McCoy in 1963. (Photo, John Green)

Dmitri Bayanov (left) and Jim McClarin at the Willow Creek Bigfoot Symposium in September 2003. They are in the Willow Creek–China Flat Museum, which has a room dedicated to bigfoot. Behind them is an enlargement of frame 350 of the Patterson/Gimlin film that was made by Marlon Davis. McClarin is holding two books authored by Bayanov. His *America's Bigfoot: Fact not Fiction* (left) is a highly detailed work on the film. Incidentally, it was Bayanov who nicknamed the creature *Patty*.

DETAILS AND ANALYSIS

1. The Decision to Ship the Films for Processing
The decision to ship the films for processing might raise some concerns. Many people would be reluctant to trust shipping such films, even if they were not totally sure they contained anything. Nevertheless, Patterson may not have really considered the full importance of the films, reasoning that as he had found one of the creatures, he could find another.

Furthermore, one will note that I am taking the stand that both films were shipped to DeAtley, not just the film showing the creature. The second film roll was available for viewing on October 26, 1967 at the University of British Columbia. It stands to reason, therefore, that it was developed at the same time and place as the first film roll. Nevertheless, Patterson could have had it developed in the interim (October 22 to 25).

2. Significance of the Plan – Film Processing
It appears evident that Patterson knew Al DeAtley could get the films developed very quickly—in one or two days at the very most. If a much longer time were needed for film development, then the men had to be prepared to stay in the area for the additional length of time. It does not appear practical that they intended waiting more than a day or so for DeAtley to get back to them (i.e., to probably telephone Hodgson's store at Willow Creek and leave a message).

3. The Decision Whereby They Both Went to Ship the Films
We can reason that it was certainly not practical for both men to travel to the shipping facility. A much better arrangement would have been for one man to travel while the other remained at the campsite to look after the horses. So why did both men travel to the shipping facility (near Eureka)? There is a very simple answer to this question. Patterson really had to go in order to make arrangements with Al DeAtley. However, they had Gimlin's truck, and Gimlin did not want Patterson to drive because he was such a poor driver. (Gimlin personally told Steenburg and me this when we were travelling to Willow Creek in September 2003).

4. The Decision to Leave their Horses Tethered at the Campsite
Both Patterson and Gimlin (especially Gimlin) were good horsemen. Being such, one would assume they were very fond of horses. We have learned that, during the excursion, Gimlin was worried about his horses back home. He wanted to get back to his animals as

The Travel Discrepancies

Al Hodgson seems to recall that when Patterson and Gimlin arrived at Willow Creek (shortly after 6:00 p.m.), Patterson stated that he had "mailed" the film to Yakima from the main post office in Eureka.

The route taken by the men as told to Hodgson was "down Highway 96 to Martins Ferry (a bridge), over the Bald Hills Road to Highway 101, and then south to Eureka (presumably).

Given this was the route taken, controversy has arisen as to whether or not one could leave the film site at about 3:30 p.m. (as Patterson and Gimlin did) and arrive at Willow Creek by 6:00 p.m. or so (i.e., in about 2.5 hours).

Daniel Perez clocked this route in 2004 and determined that it was impossible to travel the distance involved in the prescribed time, especially in 1967 when roads were probably not as good as they are now.

However, one can travel from the film site to Willow Creek in 2.5 hours, and I believe this is what the men did (i.e., went to Willow Creek first and then went to ship the film).

I have no explanation as to Al Hodgson's recollections both on the point that Patterson "mailed" the film (as opposed to air shipped), and that this had been done before he (Hodgson) met the men at his store. All I can offer is that there might be some confusion between the two occasions in which the men met with Hodgson (i.e., Willow Creek, and later the Lower Trinity Ranger Station). In other words, Patterson told Hodgson of his travel details at the latter meeting location, and other confusion has slipped in over time. (Reference: Daniel Perez, 2004. *Bigfoot Times* newsletter, November.)

In subsequent correspondence with Daniel Perez, he stated the following on the travel route issue: "Right after I got home on that trip in late October 2004, I called Gimlin and went over things. He told me he wasn't really aware of a Bald Hill Road and that as he remembers it, they went straight for Willow Creek, plain and simple, then down the 299 heading for the coast, then back to Willow Creek, then to camp, then home."

I think the second reference to "Willow Creek" is actually meant to be the Lower Trinity Ranger Station.

soon as possible. Nevertheless, as it happened, they left their horses unattended for at least seven hours in a very remote location.

I questioned Bob Gimlin on this issue and he stated that he was worried about the animals; however, the circumstances were very unusual. Both men were excited with the film and were anxious to get it to DeAtley. In other words, at that point in time, getting the film shipped was more important than the horses. I watched Bob's face very closely when he told me this, and I could see that his mind completely reverted back to the event and brought up his feelings at that time. I know he had to struggle with the decision to leave the horses. Here again, we have more evidence of the film's authenticity. I am sure Bob Gimlin would definitely not have left the horses if he knew the film were a hoax.

5. The Meeting with Al Hodgson and Contact with Al DeAtley

It is possible Patterson called DeAtley first from the pay phone and then called Al Hodgson. However, Hodgson lived only about four miles from his store, so could get there in a very short time, and the store phone would have been far more convenient for long-distance calls. Hodgson's wife, Francis, took the call from Patterson just as Al arrived home shortly after 6:00 p.m. Hodgson told Patterson to just stay at the store and went immediately with his wife and their two boys to meet him and Gimlin. They looked at a map of the Bluff Creek area and Patterson showed Hodgson the sighting location.

6. The Shipment

As to the shipment itself, it cannot be substantiated with documentation. Concerns, therefore, have been raised as to the availability of a shipping facility during the late evening. The Murray Field facility was (and is) open at all hours, and was in operation prior to 1967. Tom Steenburg and I personally asked one of the people there (a person in charge) if they would have accepted and made an immediate shipment to Yakima as late as 8:30 p.m. or 9:00 p.m. on a Friday afternoon in 1967. He said, most definitely, "that's our business." He went on to state that if there were not a pilot available (on hand), as might be the case with very late shipments, one would be called in.

I will mention at this point that there are two references stating that the shipment was sent via **mail**. However, this method is totally out of the question. The post office closed at 5:00 p.m and the men were nowhere near a postal outlet at that time. One of the references is in *The Times-Standard* article, but it should be noted that the article effectively states both air carrier and air-

Horse Sense

I have confirmed with Bob Gimlin that Patterson definitely rode a small quarter horse (which he owned), not his Welsh pony "Peanuts." Also, that Patterson had arranged to borrow a horse by the name of "Chico" from Bob Heironimus for Gimlin to use.

That Patterson and Gimlin had borrowed a horse from a man with whom they were friends, and who would later (1999) claim to have been the "creature" in the film appears odd on the surface. However, Gimlin did not have a horse that was suitable (old enough) for the expedition, so Patterson arranged to borrow Chico as stated. It is all that simple. The three men were friends and neighbors and borrowed horses from each other. Although Gimlin can't specifically recall, it is likely he had a borrowed horse for the previous Mount St. Helens expedition.

In Greg Long's book, *The Making of Bigfoot*, on page 347 Heironimus states that the reason they borrowed Chico is because it would not mind packing the "suit," and (being a mature horse) it would not be startled at seeing a man in a gorilla suit.

I think horses depend on scent mostly when it comes to other animals, so the latter point might be a little "thin." Remarkably, Chico did react when the creature was sighted (or scented), so it appears he definitely did not like something.

The other horse Patterson and Gimlin took was a little white pony that was used as the packhorse. This horse also belonged to Roger Patterson.

The Shower Story

Although Bob Gimlin cannot recall this incident, it might have some merit. John Schuchman, an acquaintance of Patterson, tells us that the two men stopped at his place in Slate Creek after they left the film site on October 21. Schuchman says the two were very excited as they related their experience. Patterson took a shower, and was within earshot of Gimlin and their host talking about the incident. At one point, Patterson jumped out the shower naked to excitedly add a comment. Schuchman personally related this story to Tony Healy. However, Daniel Perez, contacted (2007) Schuchman's daughter, Cheryl, who was not aware of the story and says that she is sure her father would have told her if it had occurred.

mail. The other reference is a statement (as discussed) made by Al Hodgson that Patterson said he had **mailed** the film from the main post office in Eureka.

7. The Call to Don Abbot and the Tracking Dogs

Calling an anthropologist in British Columbia, Canada appears strange on the surface. Why call a professional in a different country some 500 miles away? There are certainly anthropologists in California and Oregon who could have responded faster.

Don Abbott was contacted because he had previously shown interest in the bigfoot phenomenon. When the alleged bigfoot tracks were found on Blue Creek Mountain that August, Abbott flew down to the area and photographed the tracks. Blue Creek Mountain is in the Bluff Creek area. It was reasonable to assume that Abbott's involvement in the Blue Creek Mountain find would make him more likely to accept the Bluff Creek sighting. As few anthropologists anywhere gave any credibility to bigfoot, Abbott was seen as the "best bet." Contacting another professional "cold" would have been futile.

Although Abbott was specifically asked to bring down tracking dogs, the implication here is that he was to contact John Green for the provision of dogs, which he did. Green, however was unable meet this request. We can reason that the call for Abbott's involvement and for tracking dogs indicates Patterson was not involved in a fabrication. Had Abbott or any scientist gone to the film site, especially with tracking dogs, they might have easily detected a hoax. Certainly, if Patterson knew the creature was a fabrication, he would not have taken this chance.

8. The Contact with Hodgson and McCoy at the Ranger Station

When Patterson and Gimlin related the story of their experience to Hodgson and McCoy, they were still noticeably excited, or "shook-up." Patterson was walking with a limp, which he said was the result of his fall when the creature was sighted. Also, Patterson showed them a bent stirrup to confirm his account.[1] We cannot, of course, prove that Patterson's limp and the bent stirrup were the result of a fall from his horse and, as previously stated, Gimlin does not even remember this event. We might reason, however, that if Patterson planned to tell this story in a hoax scheme, he would have told Gimlin beforehand so their recollections of the sighting would match. In other words, if the event did not happen, it is unlikely Patterson would just "spring" it on Gimlin. Nevertheless, it is odd that Gimlin does not recall the fall, but such can be rationalized

Don Abbott in 1967. (Photo, J. Green)

Dogs and Bigfoot

Seen here is "White Lady" a tracking dog John Green had used in August 1967 on Blue Creek Mountain, California. Generally, we are told that dogs will not follow bigfoot tracks, and will actually cower when the creature is around or is sensed. However, this does not appear to be the case on Blue Creek Mountain.

Given White Lady or another dog (or dogs) had been brought to the film site, it would have likely followed the prints as Patterson and Gimlin did, although carried on much further. However, it started raining soon after the men called for the dogs, so I am not really sure if anything more would have been accomplished. (Photo, J. Green)

---

1. Patterson may have also previously shown Hodgson and the others the bent stirrup at the first meeting in Willow Creek.

as I have shown. Whatever the case, Hodgson and McCoy were definitely impressed with the sincerity of the two men.

9. The Contact with *The Times-Standard* Reporter
The time of the telephone call to the reporter (about 9:30 p.m.) was established from information in the article. The article states, "Patterson was still an excited man some eight hours after the experience." Given that the sighting took place at about 1:30 p.m., then it would appear the call was made at about 9:30 p.m. The contact was apparently made by telephone, and it appears Patterson did all of the talking. While Gimlin is mentioned in the subsequent news article, he is not quoted.

One might reason that the contact with *The Times-Standard* reporter was somewhat premature. Patterson was not certain he had actually caught the creature on film. He expressed this concern to Dahinden in a subsequent telephone conversation. However, Patterson did have the casts and there were two witnesses to the event. These facts, coupled with the possibility of movie footage, still made a good news story. We can conclude, therefore, that contacting a newspaper was a natural course because the men were very excited with the evidence they had gathered.

It appears *The Times-Standard* newsperson had some prior knowledge of Patterson and his activities. It does not seem reasonable that Patterson could phone in "cold," as it were, and get such good coverage of the event (see next point). Nevertheless, we might wonder why the call to the reporter was so late (9:30 p.m.). This time was over three hours after the men arrived at the Willow Creek. I am sure they were anxious to get back to their horses, so I don't think they intentionally left the call to the reporter until 9:30 p.m. It appears to me that they were unable to contact a particular reporter until that time.

10. *The Times-Standard* Article
The article was the first newspaper report on the Patterson/Gimlin experience. It appeared on the front page of *The Times-Standard,* Eureka, California on Saturday, afternoon October 21, 1967—the day after the Bluff Creek filming. Due to the importance of the information in this article, I have reprinted it here *verbatim.* I have, however, underlined certain information that I have used in my arguments.

Hodgson's store as it appeared in October 1967, although this photo was taken later.

Hodgson's store was in the same vicinity as the current Willow Creek–China Flat Museum, which Hodgson currently manages. I took these photos in 2003. The second photo was taken on the corner of the museum lot. I believe what is seen is the main intersection in the little town.

Mrs. Bigfoot Is Filmed!

A YAKIMA, WASH. man and his Indian tracking aide came out of the wilds of northern Humboldt County yesterday to breathlessly report that they had seen and taken motion pictures of "a giant hominoid creature."

In colloquial words—they have seen Bigfoot!"

Thus, the long sought answer to the validity and reality of the stories about the makers of the unusually large tracks lie in the some 20 to 30 feet of colored film taken by a man who has been eight years himself seeking the answer.

And as Roger Patterson spoke to *The Times-Standard* last night, his film was already <u>on its way by plane</u> to his hometown for processing while he was beside himself relating the chain of events.

Patterson, 34, has been eight years on the project. Last year he wrote a book, "Do Abominable Snowmen of America Really Exist?" This year he has been taking films of tracks and other evidence all over the Northwestern United States and Canada for a documentary.

He has over 50 tapes of interviews with persons who have reported these findings, and including talks with two or three persons who have reported seeing these giant creatures.

- o -

BOB GIMLIN, 36, and a quarter Apache Indian and also of Yakima, has been associated with Patterson for a year. Patterson has visited the area before and last month received word of the latest discovery of the giant footprints which have become legend.

<u>Last Saturday they arrived</u> to look for the tracks themselves and to take some films of these, riding over the mountainous terrain on horseback by day and motoring over the roads and trails by night.

Yesterday they were in the Bluff Creek area, some 65 to 70 miles north of Willow Creek, where Notice Creek comes into it. They were some two miles into a canyon where it begins to flare out.

Patterson was still an excited man <u>some eight hours after his experience</u>. His words came cascading out between gasps. He still couldn't believe what he had seen, but he is convinced he has now seen a "Bigfoot" himself

The new *Times-Standard* newspaper building in Eureka (built after 1967). The paper's logo no longer has the the symbol that replaces the hyphen. In the background of the photo we can see the old logo. What I show below the photo is from the October 20, 1967 paper.

As Patterson contacted the reporter on the same day as the filming took place, events would have been very fresh in his mind, so what was reported likely describes the incident quite accurately.

It appears Patterson telephoned the reporter, who we have now reasonably identified as Al Tostado (details are provided later under Point 10). The only mild skepticism one might harbor on the article is that Tostado did such a detailed job, and on top of that, got his article on the front page of the paper. That he appears to have quoted Patterson so much and so precisely, might be taken as a "put-up" job. Nevertheless, there were tape recorders in 1967, so he could have taped the interview and then prepared his written article. As to the front-page prominence, this was probably the editor's call. The filming was on his turf, so to speak, so why not play it up? Certainly northern California has benefited from the film through additional tourism.
(Photo, Daniel Perez)

A rewrite of the Tostado article along with my comments is provided at the following link. A scan of the original article is also provided.

http://forum.hancockhouse.com/article.php/20060612153505832

and he's the only man he's heard of who has taken pictures of the creature. Here is what he reported:

- o -

IT WAS <u>about 1:30 p.m.</u>, the daylight was good, when he and Gimlin were riding their horses over a sand bar where they had been just two days before. They had both just come around a bend when "I guess we both saw it at the same time."

"I yelled 'Bob Lookit' and there about 80 or 90 feet in front of us this giant humanoid <u>creature stood up</u>. My horse reared and fell, completely flattening a stirrup with my foot caught in it.

"My foot hurt but I couldn't think about it because I was jumping up and grabbing the reins to try to control the horse. I saw my camera in the saddle bag and grabbed it out, but I finally couldn't control the horse anymore and had to let him go."

GIMLIN was astride an older horse which is generally trialwise, but it too rared *[sic]* and had to be released, running off to join their pack horse which had broken during the initial moments of the sighting.

Patterson said the creature stood upright the entire time, reaching a height of about six and a half to seven feet and an estimated weight of between 350 and 400 pounds.

"I moved to take the pictures and told Bob to cover me. My gun was still in the scabbard. I'd grabbed the camera instead. Besides, we'd made a pact not to kill one if we saw one unless we had to."

Patterson said the creatures' *[sic]* head was much like a human's though considerably more slanted and with a large forehead and broad, wide nostrils.

"It's *[sic]* arms hung almost to its knees and when it walked, the arms swung at its sides."

- o -

PATTERSON said he is very much certain the creature was female "because when it turned towards us for a moment, I could see its breasts hanging down and they flopped when it moved."

The creature had what he described as silvery brown hair all over its body except on its face around the nose and cheeks. The hair was

WANTED

DEAD OR ALIVE
$1,000,000

Although it is common knowledge that *Life* magazine offered a $100,000 reward for a sasquatch (dead or alive), Tom Slick, the Texas Oil Millionaire, had previously offered $1 million. Cliff Kopas of Bella Coola mentioned the reward in an article he wrote for the British Columbia *Digest* magazine. He said he was informed of the offer from one of the people working with Slick on his Pacific Northwest Expeditor. I believe this person was Bob Titmus. Public knowledge of this offer, however, does not appear to have been common knowledge until July 1967 when it was disclosed in an article by T.W. Paterson entitled, "Wanted, Dead or Alive: Sasquatch" in *Real West* magazine. Unfortunately, Tom Slick had died five years earlier (October 6, 1962), so it is doubtful the offer was still valid. Nevertheless, it was an indication of what one could possibly get for a sasquatch, and Patterson and Gimlin definitely passed up a prime opportunity. To many people, I suppose, this was also a prime hoax indicator. (The image shown is a mock-up for this publication with artwork by Yvon Leclerc.)

two to four inches long and of a light tint on top with a deeper color underneath.

"She never made a sound. She wasn't hostile to us, but we don't think she was afraid of us either. She acted like she didn't want anything to do with us if she could avoid it."

Patterson said the creature had an ambling gait as it made off over the some 200 yards he had it in sight. He said he lost sight of the creature, but Gimlin caught a brief glimpse of it afterward.

"But she stunk, like did you ever let in a dog out of the rain and he smelled like he'd been rolling in something dead. Her odor didn't last long where she'd been."

- o -

LATE LAST NIGHT Patterson was anxious to return to the campsite where they had left their horses. He had been to Eureka in the afternoon to airmail his film to partner Al DeAtley in Yakima. DeAtley has helped finance Patterson's expeditions.

He and Gimlin were equally anxious to return to the primitive area. "It's right in the middle of the primitive area" for the chance to get another view and more film of the creature.

He said there's strong belief that a family of these creatures may be in the area since footprints of 17, 15 and nine inches have been reported found.

The writer jested that these sizes put him in mind of The Three Bears.

"This was no bear," Patterson said. "We have seen a lot of bears in our travels. We have seen some bears on this trip. This definitely was no bear."

Patterson is also anxious today to telephone his experience to a museum administrator who is also extremely interested in the project. "He may want to bring down some dogs. We don't have dogs here."

He's not sure how much longer they will remain in the area. "It all depends."

———

There are two points in this account that do not agree with other information we have. First, the article states that Gimlin had to release the horse he was riding. We know this is incorrect. Gimlin was able to retain his horse (he released the packhorse). Next, Patterson is said to be anxious to telephone his experience to a

Patterson's remark about the possibility of other bigfoot being in the area is likely what prompted some people to "scan" the background woods in frame 352 of the film and speculate on many shadowy images like little monkey heads peering out. The distance from the camera to the woods, however, is well over 200 feet. This distance is far too great for anything as small as a head (even a sasquatch head) to have credibility. We can hardly make out the features on Patty's face at 102 feet. At double this distance one would just see a little dark blob with absolutely no credible details.

The inset seen in the background here shows Patty at about one-half her size. I believe this would roughly be how large we would see her at about 200 feet. We can see that her head is just visible. Even if another sasquatch Patty's size was back there about 75% out in the open, identifying the image as a sasquatch would be a very tough call.

museum administrator (i.e., Don Abbott). At this point in time, Hodgson, at Patterson's request, had already telephoned Abbott. I can only conclude that the reporter did not get the facts straight in his interview with Patterson.

A question that has long puzzled researchers is the identity of *The Times-Standard* reporter who wrote the article. Recent investigation by Daniel Perez appears to provide an answer. Perez researched other *Times-Standard* newspaper articles about bigfoot that were on microfilm and found an article dated November 5, 1967 by a reporter named Al Tostado. It contained the same or similar wording and information as in the October 21 article, and in one case, an incorrectly spelled word that is found in both articles. I will also mention that the style of both articles is identical. I think it is safe to say that Tostado was the reporter.

11. The Return to the Camp Site
If Patterson and Gimlin had fabricated the film, it would have been much simpler and more practical for them to have taken their horses and gone straight to Yakima from their campsite. Their decision to leave their horses, ship the films, and later return to the campsite confirms to some degree their belief in the reality of the creature filmed. We have to ask if they would have taken this course knowing the creature was a fabrication. We can, of course, reason it was all done for effect, but this is really stretching probability.

12. Green's Message to Dahinden
The message written by the hotel desk clerk reads as follows: *"Call Al Hodson [sic], Willow Creek, Tele 916-629-2434, Very Important, happened at Willow Creek, John Green made the call 8:35 p.m., 10/20/67."*

13. The Film Processing
The processing of the films has created a lot of controversy. The type of film Patterson used required film processing under what was known as the Kodak K-12 process. It has been stated that the process was only available in San Francisco and Palo Alto, California. This being the case, it would have been impossible for Al DeAtley to get the films developed in one day.

However, I have confirmed with the Kodak people in Rochester, New York that the K-12 process was definitely available in Seattle in October 1967. I have not been able to confirm that the Kodak Technicolor Laboratory specifically had the process, but it appears it did. In any event, there is absolutely no question that Al DeAtley would have been able to get the films processed in Seattle on October 21, 1967, given a laboratory with the K-12 process was open on that day, or there was access to the laboratory's facilities.

When Patterson himself was later asked about the

Film Processing Follies

For some time it was believed that DeAtley may have taken the film to Boeing's Laboratories (Seattle, Washington) for processing, the inference here being that he knew somebody there. One would think that Boeing would have its own processing laboratory because of the sensitivity of their projects, so this was a good guess. However, when I mentioned this thought to my Kodak contact in Rochester, New York, he said no, Boeing would have used a commercial laboratory in Seattle.

I then explored the possibility of private "home" processing. I talked to a professional photographer and he said this was entirely possible. Remarkably, one professional (I believe) who saw the film stated, as I recall, that it had been overexposed and thereby indicated that it had been processed in a non-commercial lab.

For certain, the controversy over the film processing place and date will not end until this information is known and proven beyond a doubt. I would imagine the person who actually processed the film is at least "getting on" by now, and it surprises me somewhat that he (or possibly she) has not stepped forward. Surely he must get a little "kick" out of the fact that the film has become so famous.

The note left for René Dahinden at a hotel in San Francisco.

film developing by Pete Loudon of the Victoria, British Columbia *Times Colonist* newspaper, Patterson stated, **"I got them** *[the film processing]* **done at a private place. It would jeopardize the man's job if it were told."** Here, we can only conclude that the personal pronoun "I" actually means the action of getting the films processed. In other words, Patterson arranged to have the films processed privately and had DeAtley take them for processing. But then there is a problem with the rest of Patterson's statement because DeAtley cannot remember where he took the film. Surely he would remember some sort of "special" arrangement—but perhaps he does not *wish* to remember. We can reason that if the film processing would **"jeopardize the man's job,"** the man worked for a film processing facility, and had "off the record" access to film processing equipment (or had his own equipment and was "moonlighting"). Patterson and DeAtley would therefore be reluctant to name this person. This conclusion would be especially true if the man was doing a lot of "personal" or "illegal" movie film processing. Keep in mind that back in the 60s there was likely a fair demand for such processing of what are now called "adult films."

Furthermore, I have been given to believe that Patterson took a lot of movie footage, and he was certainly not beyond seeking "special" deals. I will also mention that Al DeAtley was (and continues to be) a man of "considerable means," and could certainly get things done that other people would find difficult, if not impossible.

Film Development Issue

Al DeAtley's reluctance or inability to name the film development facility or person continues to plague researchers. The information is needed to verify that the film could have been developed in fewer than 36 hours. Nevertheless, because the film has withstood intense analysis, and appears to show a natural bigfoot, it is concluded that development within the time shown did take place.

14. The First Screening of the Film Showing the Creature

I think it is important to note that René Dahinden recalled seeing film packaging material on a table when he viewed the film for the first time at Al DeAtley's home. Dahinden once emphatically stated to me, **"Why didn't I go over and look at it!"** Although this action might have answered a lot of questions on film shipping

This article that appeared on November 16, 1964 sheds a little light on the complexities of processing Ektachrome movie film. I am not sure of the application here with regard to the Kodachrome film Patterson used, but from research by Roger Knights, it was "vastly trickier" than Ektachrome. The article reads as follows:

Allprints Photo Adds New Unique Film Processing Unit

The addition of new equipment by Allprints Photo, Inc., for the processing of 16mm Ektachrome movie film has been announced by Kenneth C. Major, President.

The Mansfield installation is the only such processor unit between Philadelphia and Chicago. It makes possible overnight delivery of film taken at National Air Space laboratories in Cleveland, for example, and at North American Aviation in Columbus, and many other laboratories and industrial plants throughout Ohio.

While the installation is complicated, as indicated by the accompanying photograph, just a turn of a few valves converts the machine to handle either commercial Ektachrome or high-speed movie film.

With the addition of this new equipment, the Allprints plant becomes one of the most widely diversified photo finishing plants in America. It has broadened its field of activity into many industrial phases and is fast becoming widely known as the top finishing plant in Ohio and adjoining states.

As a licensee of Eastman Kodak, the Allprints plant was selected as one of two in the entire United States to test a new chemical system. Plants with the ultimate in control and equipment were sought to make the practical test. It proved a great success

and processing, we can still come to a reasonable conclusion on the fact that the packaging was there. If there were any irregularities associated with film shipping and processing, I am sure the packaging would have been concealed or discarded.

15. The Second Film Roll

The second film roll does not appear to have been duplicated in its entirety and it is, unfortunately, missing. I believe the screening of this roll at the University of British Columbia on October 26, 1967 was the first and last major screening. At some point in time, Patterson gave René Dahinden about 10 feet of film from this second roll. René had five frames made into still photographs, two of which showed Patterson holding footprint casts at the film site.

Then in 1996, a short film segment consisting of sixty frames showing four of the creature's footprints in a series appeared on the Internet. I do not know the source. I believe these same shots were used in a 1975 BBC television documentary series entitled *Fabulous Animals*. This series, featuring the noted naturalist Sir David Attenborough, included a segment on bigfoot.

Mrs. Patterson has told me that the film was loaned to the BBC at or about that time and was not returned. The *Fabulous Animals* series was never made available commercially, so I am unable to determine exactly what film frames were used. It is certainly possible this was somehow the source of the frames that appeared on the Internet.

Furthermore, a BBC bigfoot television documentary, *The World's Greatest Hoaxes*, released in 1998, shows shots from the second roll (Patterson making a cast at the film site). I am positive Dahinden did not provide his 10-foot strip to the BBC. Requests to the BBC to look for the film roll did not receive a response.

Nevertheless, as they say, "wonders never cease." Upon visiting John Green in 2002, he showed me a video of the Patterson/Gimlin film along with other material on the same tape. Surprisingly, one of the still images Dahinden had made (one of two showing Patterson holding casts) came up. John could not recall how he obtained this image. For certain, Dahinden had not given it to him.

When I again visited John Green in May 2003 with Vancouver Museum people, John showed us the actual 16mm Patterson film of the creature. After the film was finished, John inadvertently let the entire film roll expire (i.e., go right to the end of he roll). All of a sudden, the footage of the creature's tracks in a series appeared. John was as surprised as I was. He did not know the footage was there. Whenever he views the

and takes much of the human element out of the mixing of chemicals for Kodachrome processing.

In releasing the news of this expansion in Allprints' equipment and operations, Mr. Major revealed that his company will continue to expand its color work for both professionals and amateurs. Its well established daily pick-up and delivery services for one-day processing of amateur film for its network of Ohio dealers will continue, as will its overnight processing of commercial and industrial 16mm film. Mr. Major also reported that Allprints Photo, Inc., already established as one of Mansfield's staunch and growing companies, is enjoying a good year.

(Photo Caption): MORE NEW COLOR FILM PROCESSING EQUIPMENT has been installed by Allprints Photo, Inc., at its plant at 174 South Main Street, already one of the most modern in the entire nation. The new equipment further improves the quality and speeds the delivery of finished film for its ever-growing list of commercial, industrial, institutional and governmental customers. Kenneth C. Major, President of the firm, is shown with some of the intricate plumbing that makes up a part of the new installation.
(Source: Scott McClean collection.)

Comment: If this information has applicability to the Patterson/Gimlin film, I think we can reason that simply processing in a home laboratory would be extremely difficult. The use of such equipment "after hours," therefore appears to be more plausible.

David Attenborough also wrote a book on the *Fabulous Animals* series. As to his opinion of the Patterson/Gimlin film, from what I can gather, he states that he thinks Patterson was hoaxed. It would be interesting to see exactly how he came to that conclusion.

film, he shuts off the projector at the end of the main footage (i.e., end of film). He has no idea how the additional footage got on the roll. Somewhere down the line, it appears Al DeAtley or someone spliced the footage of the creature's prints in a series. There are a number of people who have a copy of the film, so perhaps their copy also has the additional footage, and that could also be how it found its way to the Internet.

At one point, prior to Patterson's death, Gimlin inquired of him as to the location of the second roll. Patterson informed Gimlin that Al DeAtley had it "somewhere." Gimlin followed up with DeAtley, but he denied both having the film and that it ever existed. Evidently, DeAtley forgot that he probably had the film developed.

While working with Yvon Leclerc on my book *Meet the Sasquatch* (Hancock House, 2004), Yvon assembled (registered) the sixty frames from the Internet into a single image. I recently (March 2005) discovered that the last footprint in the series shown matches the shot of a single footprint from the 10-foot strip. This was not previously noticed because the print is mirror-imaged (film lab error). This discovery leads me to believe that the 10-foot strip also contains all of the other footprints. Furthermore, the total number of 16mm frames on 10 feet of film would be about 360 frames (about 23 seconds of viewing), so there are likely other important frames on this strip.

16. The Original Film Roll

Unfortunately, the original film is not available for inspection. While its availability is not essential for any viewing purpose (there is a master copy and very good duplicates) it would be highly desirable to confirm the film processing date which may be shown on the film leader or trailer. Generally, but not necessarily, film processing facilities showed the processing date. The last word on the original film is that it was loaned to American National Enterprises, and later got caught up in a mass of litigation, brought about for the most part by René Dahinden (discussed in the next section). It subsequently ended up at a law firm in Florida. I believe the film could be recovered. It is just a matter of time, money, and dealing with some legal entanglements that go beyond the scope of this discussion.

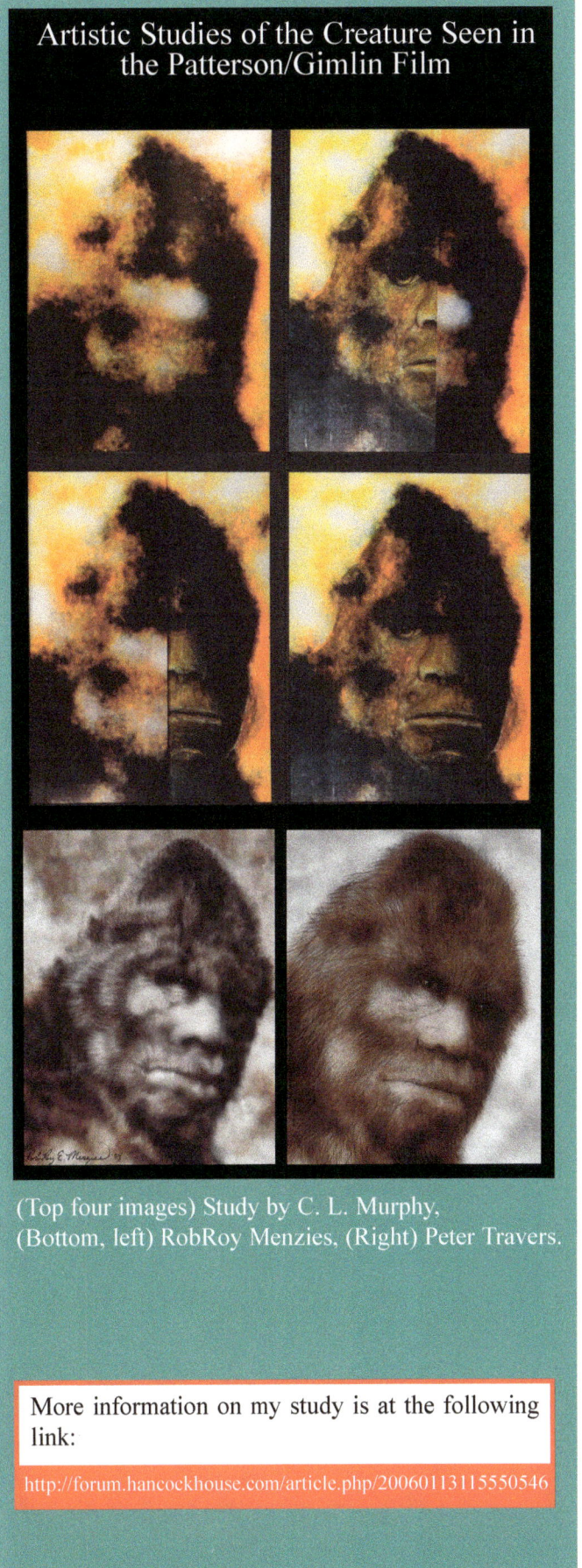

Artistic Studies of the Creature Seen in the Patterson/Gimlin Film

(Top four images) Study by C. L. Murphy,
(Bottom, left) RobRoy Menzies, (Right) Peter Travers.

More information on my study is at the following link:

http://forum.hancockhouse.com/article.php/20060113115550546

6. Credibility of the Source

Patterson and Gimlin's bigfoot sighting experience at Bluff Creek was unique for three main reasons. First, there were two witnesses to the event, so their stories could be cross-checked. Second, there was photographic evidence in the form of the movie Patterson took which could be analyzed. Third, the men had plaster casts of the footprints made by the creature to further confirm the sighting. Moreover, many of the footprints at the film site remained intact and could be seen, photographed, and cast by other researchers. All of this evidence was certainly very credible. Important questions, however, remained unanswered on the credibility of Patterson and Gimlin themselves. These questions were pursued by René Dahinden and John Green.

Dahinden states that he attacked the issue from the negative perspective. In other words, he looked for evidence that the film was a hoax. The backgrounds of both Patterson and Gimlin were researched. Patterson was a family man. He had a wife and three children to support. He did not have steady employment and basically made a living any way he could. Work in rodeos provided some of his income, and he had invented a number of useful devices.[1]

One device, called a "Prop Lock," was a plastic holder used on fruit tree props. This device, which was like the clip used on bread bags, held branches firmly in place. It eliminated slippage during moderate or heavy winds—a common occurrence with the standard 'Y' props.

One of Patterson's former business partners stated that he felt Roger could have made "good money" on some of his ideas. Patterson, however, did not take the time to fully develop and promote his inventions. It appears he became bored or impatient and went on to new projects. Financially, Patterson was often in need of money.[2] He had borrowed money and assumed other debt, and was delinquent in his payments. No one, however, considered Patterson *basically* dishonest. He was a very likable, easygoing person who appeared to have good intentions.

Patterson's book, *Do Abominable Snowmen of America Really Exist?* (Franklin Press, 1966) is primarily a collection of bigfoot-related newspaper and magazine articles.[3] It includes a number of drawings by Patterson that illustrate the published bigfoot incidents. There is no doubt that Patterson had above average artistic abilities.[4] To his credit, Patterson did "get it together" and write/compile a book, providing excellent references for that period in time. On October 27, 1966,

Roger Patterson is seen here in 1967 with one of his Welsh ponies. An ad for his "Prop Lock" invention is shown atop his van. (Photo, J. Green)

In this photo, also taken in 1967, René Dahinden is seen on the left with what appears to be a full-grown Welsh pony. Roger Patterson is seen on the right with what is probably a Welsh pony colt. Dahinden told me that Patterson could transport two ponies in his van. Judging by the size of the full-grown pony seen here, I would image they would have been a little cramped for head room. (Photo, J. Green)

Bob Gimlin's home, Yakima County, Washington. The photo was taken in 1996.

Patterson sold the book copyright to Glen Koelling for the sum of $500. At the back of the first edition of the book (second printing), there is an invitation to join the *Abominable Snowman Club of America*. This club was formed by Patterson and Koelling. Later, when Patterson formed the Northwest Research Association, he also called for memberships.[5]

Shortly after Patterson published his book, he posted a $100,000 reward for the capture of a bigfoot creature. The reward was made in the name of his company, Trail Blazers Research Institute, Inc. Patterson stated that in order for a person to collect the reward, the creature must be alive and in good condition. I do not know what backing he had, if any, to provide the reward.

Healthwise, fate had dealt Patterson a devastating blow. He had Hodgkin's disease and knew he would die a relatively young man. People who knew him offer that he became *less* responsible upon learning of his disease.[6]

Dahinden and Green were fully aware of Patterson's "happy-go-lucky" reputation. They researched all aspects of his personal life as far as they possibly could. In Dahinden's words, *"We took him right down to the underwear, because we didn't want to be fooled."* Every effort was made to find out if Patterson had been involved with anyone associated with costume making or theatrics. All rumors or hearsay remarks that implied a hoax were checked out. Neither man uncovered information that connected Patterson with a hoax. They certainly found out that Patterson had a less than "solid" reputation. Yakima publications refused to carry the story of the film for this specific reason. At one point, Patterson confided in René Dahinden, "I am the worst person that this *[the bigfoot encounter]* could have happened to."

Some information surfaced on April 4, 1968, which has led to unjustified and highly improbable speculation. On this date, the *Yakima Herald Republic* reported that George W. Radford, Jr., and his wife Vilma had filed a suit against Patterson for failure to meet a financial commitment. Patterson had borrowed $700 from the Radfords on May 26, 1967. He promised to repay them $850 on June 10, 1967, plus give them 5% of the proceeds from his *planned* bigfoot documentary. He defaulted on both accounts. In the suit, the Radfords claimed the agreed sum ($850), 5% of the Patterson/Gimlin film proceeds (film obtained October 20, 1967), together with costs and a reasonable attorney's fee. Actual court papers (Case #51297, March 1968) confirm that this information is correct.[7 and 11]

While the final outcome of this case (as detailed) is

Roger Patterson, left, is seen here with Dennis Jensen (who was employed by Patterson), viewing the countryside (a canyon) near Oroville, California, in 1969. Jensen had found footprints in the area and Roger went to investigate. One thing I think we can determine is that Patterson was a pretty steady field researcher. He definitely got "out there" and looked for bigfoot. This fact that he filmed one, therefore, should not be considered too unusual. Thousands of people claim they have seen the creature. Patterson just happened to be better prepared and managed to get about one minute of film footage.

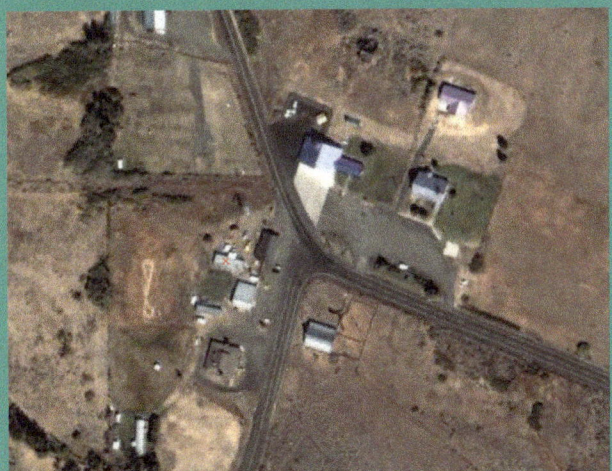

At the time of Roger's death, the Pattersons lived in Tampico, Washington at this location. After Roger died, their house was taken down and replaced with a portable building, indicated in this image with a red "x." Patricia Patterson lived here until she recently moved to Yakima. In 1967, the family lived about 7 miles southwest of this location on the same road. As can be seen, this is a very rural area. Beyond the scattered farms, there are vast mountainous forest regions. (Photo: Image from Google Earth. © 2008 Digital Globe)

amusing, we can draw nothing from it except Patterson's reluctance to repay his debts. If the money was to be used to perpetrate a hoax, as some people have speculated, the amount of $700 does not appear to be anywhere near adequate. Certainly, if there were a number of contributors involved besides the Radfords, each providing a like or greater amount, there is some room for speculation. However, no other similar claims against Patterson were made, and no one else has come forward stating money was loaned to him. Whatever the situation, considering his obviously poor credit rating, Patterson had to present a pretty good story to the Radfords to get the loan. In all likelihood, he sold the Radfords with his enthusiasm for bigfoot. The fact that he promised to repay the loan in the short time of about two weeks would imply that he had some fast money-making scheme in mind, but certainly not a bigfoot hoax scheme, other than to use the money to get more investors.

Bob Gimlin was an open and shut case. Gimlin was, and continues to be, a responsible, sensible, honest person. He has been married twice and has four children. He served in the U.S. Navy during the Korean War. He has served with the Sheriff's posse and also the Search and Rescue Team in Yakima County. Partly of First Nations heritage, Gimlin is an easy-going and respectable person. Dahinden and Green quickly concluded that Gimlin was totally above suspicion regarding a possible hoax.[8]

Both Patterson and Gimlin's sincerity with the film impressed the researchers. Patterson was excited and full of emotion. Gimlin, although less outward in his reactions, shared Patterson's enthusiasm.

Although somewhat suspicious of Patterson, most bigfoot researchers gave him the benefit of the doubt. The film of the creature was very convincing, and Gimlin was beyond reproach. Investigation of the circumstances related to the film therefore trailed off. It was replaced by more intensive examination of the film itself. The rationale in this regard was that if the film were a hoax, the film itself would reveal this information.

One major source that, to my knowledge, was not subjected to any intense investigation was Al DeAtley (officially, Al DeAtley Jr.), Patterson's brother-in-law. DeAtley was (and is) a man of "means" who helped Patterson financially with his business pursuits and bigfoot expeditions. DeAtley was in partnership with Patterson and Gimlin. The three men formed a business association on November 1, 1967, whereby each man held one-third ownership in the bigfoot film. The business name of their company was Bigfoot Enterprises. Initially, DeAtley was highly interested and active in

Bob Gimlin in 2003.

In the last few years or so, Bob Gimlin has become much more open about his experience at Bluff Creek. He has attended several Bigfoot symposiums and given talks. He has also done several video interviews. He is so down-to-earth and straightforward in all aspects of h is life, few people doubt his word.

the new company, and worked with Patterson to commercially exploit the film. It is apparent, however, that DeAtley lost interest because, sometime prior to August 5, 1970, he transferred his film rights to Roger Patterson, giving him two-thirds interest.[9] One thing we need to keep in mind is that DeAtley was an educated businessman and certainly not a person who could be easily fooled or coerced. It is highly unlikely Patterson would try to "put one over" on this man, and equally unlikely that DeAtley would agree to be part of a hoax scheme.[10]

When Patterson died in January 1972, his widow, Patricia Patterson, was vested with her husband's share in the film. How much money was made on the film up to this time is not known, but it appears there was very little left.

The following is reprinted from Dr. Grover Krantz's book, *Bigfoot/Sasquatch Evidence* (Hancock House, 1999) p. 120:

> A few years after the film was made, Patterson received a letter from a man in Thailand who assured him a sasquatch was being held captive in a Buddhist monastery. Patterson spent most of his remaining money preparing an expedition to retrieve the creature. I was to be part of the operation. Then a man who was sent to investigate on the spot found out that it was a hoax. At the time, Patterson knew he was dying of Hodgkin's disease and firmly believed that with enough money he might be cured. Instead of making another Bigfoot movie which he could have done if he had faked the first one, he spent almost everything he had on a wild goose chase. Then he died.

I do not believe the film proceeds since 1972 have been "significant." I say this because I do not believe Mrs. Patterson has ever been financially "well off." I visited her in August 2003 and can attest that her lifestyle is reasonable.

What is known, however, is that Gimlin did not receive any compensation from the film. Here, the general speculation has it that Patterson and DeAtley cheated Gimlin. This speculation is not totally correct. The truth is that Gimlin had problems with film publicity (personal appearances). Also, he and his family had been subjected to some ridicule as a result of the film. Bob thereupon ceased to make personal appearances with Patterson and DeAtley on their "road tours" showing the film.

As a result of this action, Patterson hired a "stand-in" for Gimlin (certainly a "shady move," but that's

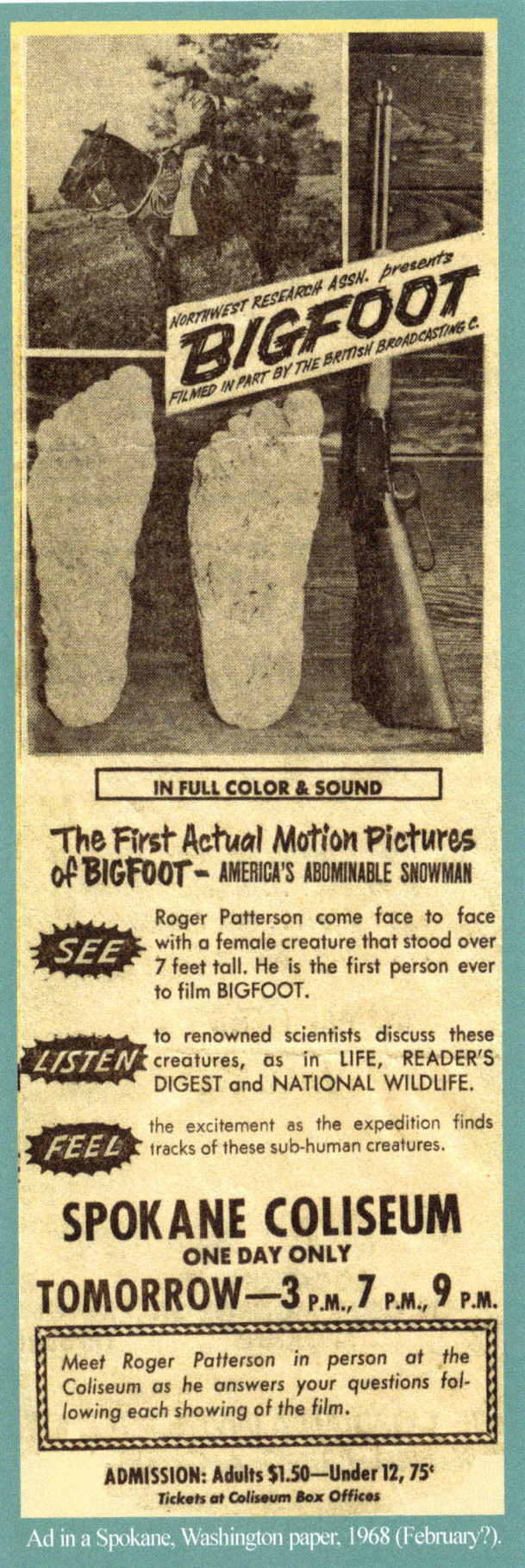

Ad in a Spokane, Washington paper, 1968 (February?).

show business). In time Gimlin specifically requested that he be disassociated from the company in all respects. While Patterson and DeAtley were not justified in failing to provide *some* compensation to Gimlin, it appears they felt justified because Gimlin did not wish to work with them, although nothing was put in writing.

In 1974, however, René Dahinden persuaded Gimlin to change his mind. Dahinden offered to hire an attorney and sue the Patterson estate for Gimlin's rightful compensation. In return, Dahinden wished to become an equal partner with Gimlin in his share of the film. Gimlin agreed, and the issue was eventually resolved in an out-of-court settlement on February 6, 1976. Under the settlement, Gimlin and his wife agreed to forgo any portion of the past film proceeds collected by Patterson and DeAtley (now just Mrs. Patterson) in return for 51% ownership in the film, plus exclusive film publication rights for selected film frames (i.e., printed material rights). Dahinden thereupon initiated many lawsuits against companies that had used film material without compensating Gimlin. Dahinden also undertook a massive "witch hunt" and found many instances in which film material had been used in movies and books with absolutely no approval and no compensation to anyone. The mass of legalities started to intrude on Gimlin's private life and by late 1978, he had had enough. Gimlin turned over (virtually gifted) his film rights to Dahinden on September 29 of that year.[11]

Another person who I thought might have been able to shed a little light is Roger's brother, Les Patterson. He owns and operates Patterson Rodeo & Livestock Company in Pasco, Washington (about 80 miles from Yakima). Les was definitely not interviewed by John Green. When I mentioned Les to René Dahinden, he said he knew of him, but that was all. So I am certain René did not interview him either.

I talked with Les on July 27, 2003. He was highly supportive of Roger and is a firm believer that the film is genuine. He told me that he and Roger went to the Bluff Creek area after the filming on an expedition sponsored by the Ford Motor Company and J.B. Hunt (now owner of J.B. Hunt Transport Inc.). Les provided ten horses for the expedition. He recalled that people in Humboldt County treated Roger like a hero. The expedition failed to find a bigfoot, but did find tracks. Les further stated that even on his death bed Roger did not waver as to the authenticity of the film.

There are two other Patterson brothers, Lorn and Glen, both still living at this writing. I have not contacted these brothers.

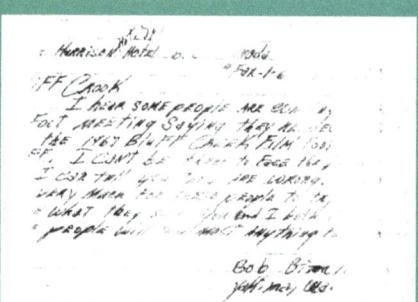

Bob Gimlin has been continually under pressure as to his insistence that the bigfoot filmed was a natural creature. In 1996 a bigfoot symposium was held at Harrison Hot Springs. Bob could not be there, and was greatly concerned about something he had heard, so he sent the above message to Cliff Crook, who was in attendance. The message speaks for itself. It reads:

Harrison Hot Springs Hotel, B.C. Canada
[July 22, 1996 - not shown on scan]

Cliff Crook

I hear some people are coming to the Bigfoot meeting saying they helped Roger fake the 1967 Bluff Creek film footage. Cliff, I can't be there to face the people but I can tell you they are wrong. I would like very much for these people to try and prove what they say. You and I both know some people will say most anything for a price.

Bob Gimlin
Yakima, Wa.

I was at that symposium, and as I recall whoever the people were, they did not show up.

John Green and René Dahinden bought the Canadian screening rights for the Patterson/Gimlin film in early 1968. It was presented in various B.C. towns for a year or so.

DETAILS AND ANALYSIS

1. Patterson's Skill in Inventing

We can certainly speculate that Patterson could have invented some device or process for fabricating excellent footprint casts. Mechanically fabricating the actual footprints in the soil at Bluff Creek, however, would have been much more difficult. The footprints indicate they were made by a flexible foot. They were also deeply set in the soil, which would require weight or pressure of some sort. These conditions do not imply that the task was impossible, just highly difficult. Moreover, fabrication of the casts or the actual prints could not have reasonably occurred without Gimlin's knowledge.

2. Patterson's Financial Situation

Patterson poor financial situation probably presents the strongest argument that Patterson engineered a hoax. He needed money badly. Fabricating a bigfoot film would not constitute a crime. Claiming that the creature was real and marketing the film under this pretext might result in fraud charges **if he were caught.** The penalty, however, would not be very severe. The fact that Patterson had Hodgkin's disease would be taken into consideration. The odds for success were definitely on his side. Providing conclusive evidence that the film was a hoax would be both time-consuming and costly. In this connection, it would be up to the **victims** (if there were any) to prove the film was hoaxed. It would not be up to Patterson to prove the film was real. In my opinion, however, Patterson simply thought that he could make some money with a bigfoot film documentary and this was the full extent of any "financial motivation."

3. Book Aspects

The full history of the book is as follows. Patterson had the book printed in 1966 and showed it to Glen Koelling. This was the original book, or more appropriately it was an unbound manuscript. I don't know how many copies were printed, but doubt that there were very many. Koelling bought the book copyright for $500 that same year (agreement is shown on the right). He cleaned up the work, finished it, and then had it printed by Franklin Press, producing the first edition. The copyright is shown as Trailblazer Research, Inc., of which Koelling was the president. Patterson then made a few minor changes (cover, title page, new photo of himself) and had the book reprinted as the second edition (probably 1967—no date is shown). He shows himself as the copyright holder, and the publisher as Northwest Research Association (a company he formed in or about 1966). René Dahinden purchased the copyright from Koelling in 1979 for $500.

1966

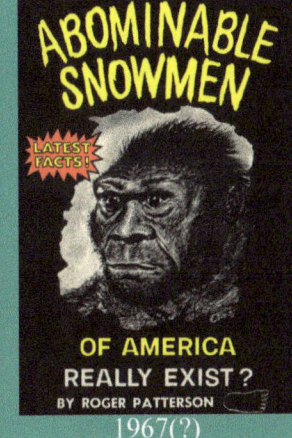
1967(?)

1996

2005

I republished the book under Pyramid Publications (my company) with Dahinden's permission in 1996. I listed Koelling as a copyright holder (1966) at Dahinden's direction. In 2004, I added an update supplement to the work, plus some other material, and it was republished by Hancock House Publishers in 2005 under the title, *The Bigfoot Film Controversy*. The covers of the book editions are shown here in order.

Koelling did not mention Patterson's "second edition" when he was interviewed by Greg Long (*The Making of Bigfoot*, pp. 196-207). However, Koelling told me he heard about it, but was too involved in other things to make an issue of it, and I do believe close friendship was involved here.

On November 23, 2005 Glen Koelling responded to an article I posted to my blog, The Murphy File, (Hancock House Topic Forum) regarding Patterson's book. He was upset with the misleading information given in Greg Long's book, *The Making of Bigfoot*. Koelling stated the following:

> My name is Glen Koelling, the individual quoted in several books as the person who helped Roger Patterson publish his first book, *Do Abominable Snowmen of America Really Exist?*. It is amazing to me the false statements, innuendos, and downright lies concerning Roger, me and the book. I had the greatest admiration for Roger, never felt that I was lied to, he did not owe me money and to this day believe his film was not a hoax. Bob Gimlin was a man of integrity, loyal, and a good outdoorsman. I am appreciative of the opportunity I had to be associated with both of these men, had a great experience in publishing the book and working my tail end off getting it on the market, and meeting people who had seen bigfoot. Too bad some self-serving authors have to distort the truth and malign someone who had the guts to get out there and do something besides write about it.

We can reason that Patterson compiled the book to draw attention to the bigfoot phenomenon and make money. Did he compile his book as a prelude to a major hoax? It is indeed coincidental that the book was published in 1966 and the epic bigfoot encounter took place the following year. A connection is possible, but really far too obvious.

4. Patterson's Drawings
Patterson's drawings indicate that he had good perception. Three of his drawings show female bigfoot creatures. Two drawings are his, the other is a copied draw-

---

A Little Tribute to Roger Patterson as Provided to Me by Glen Koelling November 25, 2005

"He made mistakes as any human will do, but I can honestly say that I never knew him to knowingly attempt to mislead, misuse, cheat or take advantage of anyone. He was driven, to those such as he who succeed we forget any and all of the negatives, to those who don't quite get there, such as Roger, the critics vilify, and to me this is sad. To some extent, the failure can be laid at my feet because I had difficulty staying focused on what we were trying to accomplish and I believe, had I done so, things would have turned out differently."

ROGER C PATTERSON
WASHINGTON
PVT US ARMY
KOREA
FEB 14 1933   JAN 15 1972

Roger Patterson rests in the West Hills Memorial Park, Yakima, Washington. Shown here is the information on his grave stone. Funeral Services were held in the Shaw and Sons Chapel, Wednesday, January 19, 1972. Ron Olson and Bob Gimlin were two of the bearers. The memorial card for his funeral carries the little message, "On the wings of today comes our strength for tomorrow."

A rose that bloomed in January.

ing. From this evidence, we know he thought about gender in connection with the creature. It is, therefore, feasible that he would consider a female form if planning a hoax. His main reason for this (as previously stated) would be that a female would be given more credibility than a male. Nevertheless, to consider this highly marginal evidence as an indication of a hoax scheme is not practical.

5. The Memberships
According to Glen Koelling, Patterson did not meet his commitment to provide a club newsletter for the Abominable Snowman Club. Koelling tells us he returned the club fee of $5.95 plus 50 cents postage/handling to the applicants. When Patterson later founded the Northwest Research Association, he did issue bulletins, although I don't know how many. He also put together a 68-page booklet in 1968 containing sightings, articles, and letters regarding the film and other subjects. Most of the pre-film material is simply a reprint of that used in his book, *Do Abominable Snowmen of America Really Exist?* The booklet was sold, and in it he mentions he was working on a second volume that he offered for $4.00.

6. Patterson's Health
The fact that Patterson had Hodgkin's disease might be considered a strong motive to perpetrate a moneymaking hoax. He had a wife and children and was probably very worried about them. It is certain he had no savings or investments to provide for his family after his death. There is also the possibility that his impending death urged him to seek some way to make a mark in life. As he had been basically unsuccessful in his business endeavors, he might naturally have dreamed of some "grand exit." The bigfoot field was wide open and primed for a major sighting. Like all bigfoot researchers, he felt discovery of the creature itself or its remains was just a matter of time. He may have reasoned that a hoax would not physically harm anybody. However, this kind of speculation gets tiresome, mainly because it is so "convenient." We must remember that Bob Gimlin, Al DeAtley and possibly others had to be involved in a hoax scheme, and they all had to make a living. Think about this a little. What is your opinion of people who willfully perpetrate hoaxes? Would you hire such people or contract services from them? Hoaxes are considered fun in some circles, but the business world is far less forgiving.

7. The Radford Case
In response to the Radford's claim, Patterson provided a Promissory Note on January 7, 1969 whereby he

Membership certificates issued by Roger Patterson for his club and association.

Booklet by Patterson, 1968.

A synopsis on this booklet is provided at the following link:

http://forum.hancockhouse.com/article.php/20060313231402708

agreed to pay them $400 immediately and $700 by April 1, 1969 (total of $1,100). Patterson failed to pay the $700, so the matter again went to court (Case #23927, January 6, 1970). Patterson was ordered to pay the Radfords $999.99, which constituted his entire debt. In other words, no further claim could be made against him (the 5% film rights were forfeited).

On June 29, 1970 an Order of Dismissal for Want of Prosecution was signed and issued on the original court case (#51297). I do not have a similar document for the second case (#23927). I do not know if Patterson met his commitment. It appears the Order for Dismissal on the original case might also have applied to the second case. If so, then we can reason Patterson did not make the payment ($999.99) ordered by the court.

Apparently unaware that they were no longer entitled to the 5% film rights, the Radfords sold this right to René Dahinden for $1,000 on November 16, 1974. He gave them $100 cash and promised to pay them the additional $900 when he, "collected from Mrs. Patterson."

On January 25, 1977, Dahinden filed a claim against Mrs. Patterson for payment of 5% of the film proceeds. Mrs. Patterson (now holding all the cards) did not respond. On March 3, 1977, Dahinden requested the court for an order adjudging Mrs. Patterson to answer his claim. On October 21, 1977, Dahinden received an answer which was as follows: "On January 6, 1970, judgment was granted in the sum of $999.99. Thereupon plaintiff's *(Dahinden's)* instant claim was merged into said judgment causing a bar to further prosecution of the said claim." On August 18, 1978, Dahinden's claim against Mrs. Patterson was dismissed. The lawyers for both Dahinden and Mrs. Patterson signed the dismissal.

8. Gimlin's Influence, Actions, and Position

There can be no doubt that Bob Gimlin's exceedingly high credibility overshadowed suspicions relative to Patterson. This fact holds true to this day. If Gimlin, or another person with equally high standards, had not been present at the filming, initial belief in the film would have been greatly diminished. This condition would have also led to considerably more skepticism on the part of the other researchers.

In the course of events, which span over 39 years, Gimlin has never wavered in his testimony. He currently does not have any legal rights to the film. His rights are now held by Erik and Martin Dahinden, René's sons. Gimlin does not receive any financial proceeds whatsoever from the film.

Gimlin virtually divorced himself from the incident

The Paper Jungle

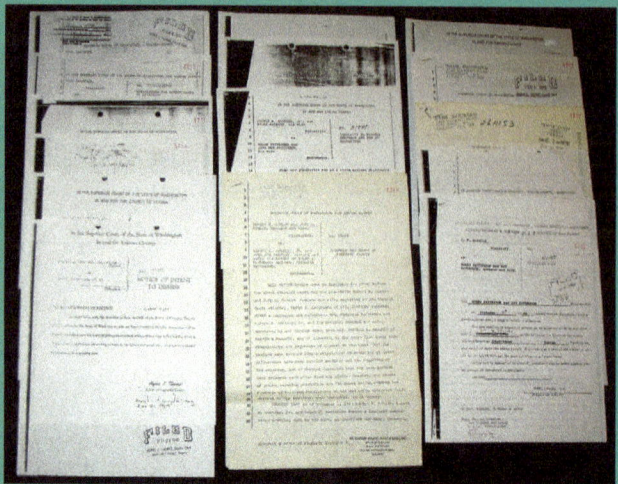

Without doubt, there is a mass of court/legal documents related to Patterson himself and the film. They were provided to me by both René Dahinden and Cliff Crook. I don't have them all, but certainly the most important ones. I have studied them, and summarized them. I even scanned them and considered presenting them in this work. I came to the conclusion, however, that what they say is totally irrelevant to the issue of whether or not Patterson and Gimlin filmed a natural bigfoot. The notion that because Patterson had debts and failed to return a rented movie camera is hardly an indication of a hoax. Nor is all the haggling that took place over the film copyright issue. It surprises me a little that if the film were a hoax, that someone "in the know" did not step forward. I am sure the word got around Yakima that the cases were being heard.

> For those of you who wish to delve into the legalities, I have provided the documents in my possession at the following link:
>
> http://forum.hancockhouse.com/media/legal%20documents.new.pdf

TEAM ROPING   CONTEST HORSES   BULLDOGGING

PATTERSON RODEO & LIVESTOCK CO.
SPECIALIZING IN CORRIENTE CATTLE

LES PATTERSON, OWNER   WEST 9355 SAGEMOOR ROAD
(509) 547-9978   PASCO, WA 99301

Les Patterson's business card (Roger's brother).

after giving his film rights to René Dahinden. Until the last few years or so, he did not normally allow interviews (even paid interviews), nor normally accept invitations for personal appearances.

A telephone interview (1997) with Gimlin for the television documentary **Shooting the Bigfoot** was an exception. In this interview Gimlin was asked if a hoax was possible. Gimlin replied that although he is still convinced the creature was real, a hoax was possible. The documentary concluded that the film was a hoax. Certainly, almost any question on "possibility" has to be answered in the affirmative. The question should have been on probability, not possibility. Without doubt, the question was an intentional media ploy—one of the cheap tricks used by disreputable media people to create controversy.

9. DeAtley and His Share of the Film
The main question in this regard is why did DeAtley surrender his share of the film rights? In my opinion, DeAtley simply considered the film more trouble than what it was worth. He was financially independent and really did not need the stress associated with the film. Now, having said that, I am told DeAtley used the money he made on the film to invest in his company, which was not doing that well at the time. The money did enable him to "turn things around." We have lately learned (Long, *The Making of Bigfoot*, p. 215) that DeAley said he gave his film share to Roger Patterson because he had cancer and needed the money. I suppose one could "make hay" out of all of this, but any speculation is really far beyond reason.

10. DeAtley's Knowledge
We can conclude that DeAtley knew Roger Patterson very well. DeAtley is married to Roger's sister, Iva, and as previously stated, had assisted Roger financially with his business pursuits and bigfoot expeditions. We can therefore assume that DeAtley had the "inside track" on a lot of Patterson's initiatives. However, the only meaningful insight I can gather from information on DeAtley is that he says he does not have any proof that the film is a hoax. Whether he would really tell us if he did have proof is, of course, another matter. Nevertheless, I personally think that DeAtley simply did not see or hear anything that was "out of line." That does not mean that such things did not occur, they just surprisingly got past someone "sitting in the front row." By his own admission, DeAtley stated he did not want to know all the details. He did not believe in the existence of bigfoot and let that fact stand as to his opinion of the film. However, when professional people gave credibility to the film, he admits that he changed

Roger Patterson is seen here (right) with René Dahinden at Patterson's home in the spring of 1967. According to Greg Long (*The Making of Bigfoot*, 2004), Patterson had been involved in a lot of "questionable" bigfoot-related activities right up to the filming of the creature at Bluff Creek in October 1967. Despite what people say about Dahinden's unusual character and lack of diplomacy, I can say with certainty that he had his "ear to the ground" on everything involving bigfoot and was not easily fooled. During my numerous discussion with René, he never told me of anything that indicated Patterson (and/or Gimlin) fabricated the film. (Photo, J. Green)

I do not have a photograph of Al DeAtley and have never seen one. I note that Greg Long does not show a photo of him in his book, so strongly suspect that DeAtley would not allow him to take one. He is definitely the "mystery man" in the Patterson/Gimlin saga, but I think that if he knew or remembered anything more than he has stated, we would have learned of it by now.

his mind. From this we can reasonably conclude that DeAtley definitely does not *know* anything about an unlikely fabrication. Knowing Roger as he did, DeAtley does offer that Roger was too smart to include a third person in any hoax scheme (i.e., someone besides Gimlin wearing a "suit").

11. The Legal Proceedings
The mass of legalities on film ownership and usage is really a side issue. While marginally interesting, the underlying tone is that everyone wanted to make money on what they thought was, or could be, a film of a natural bigfoot. If the question is, did Patterson exploit the film to make money? The answer is yes. If the next question is, was everything done legally? I would have to answer yes, as I don't know of anyone who went to jail or who was successfully sued. In the lawsuits, each side thought they were right, and judges made the decisions, or out-of-court settlements resolved the issues. The only marginal inference some people might gather from Patterson's actions is that he was not totally truthful in all of his dealings. Therefore, he may not have been truthful about filming an actual bigfoot. This is a consideration, I suppose, but hardly a reasonable measure of proof that the film was fabricated.

Bob Gimlin is seen here (left) with me and my film site model at the Willow Creek Bigfoot Symposium in September 2003. Gimlin is an extremely personable man, and so far removed from what one might deem a "con man," or someone capable of pulling off a major hoax, that the idea is, in my opinion, ridiculous. Could Gimlin have been fooled into thinking a man in a costume was a bigfoot? I really don't think so. At 100 feet or less, neither he, you, or I would be fooled, especially back in 1967. (Photo, Thomas Steenburg)

The Third Man

Even the lowest estimate of the creature's height rules out Patterson or Gimlin as playing the part. I personally believe Jeff Glickman (a certified forensic examiner) was correct in his conclusion that the creature was 7 feet 3.5 inches tall, which rules out most of humanity.

Here Comes the Judge

The thought has often crossed by mind, and probably the minds of others, as to the outcome of a court case conducted on the existence of the sasquatch. The Patterson/Gimlin film would certainly become Exhibit "A" for the prosecution, followed by a stream of other evidence such as footprints, casts, and so forth. The defense (scientific establishment) would have little or nothing to go on save a mound of hearsay evidence (with Greg Long's book topping the pile), and a statement of scientific protocol: "It cannot be so, therefore it is not so." Nevertheless, the rule of *habeas corpus* would likely turn the tide, but it would still be a tough call, and many scientists would probably give the whole issue much more thought.

7. The Scientific Community and the Press

John Green and René Dahinden convinced Patterson that the scientific community in British Columbia, Canada would probably be the most receptive to the new film evidence. This advice was based on the fact that people in British Columbia were very familiar with the bigfoot or sasquatch phenomenon. Green and Dahinden had been "cultivating" this region with their research for many years in hopes of gaining government backing for their endeavors.[1] Patterson was further persuaded by a telephone call from Don Abbott, who promised good scientific representation for a viewing of the film. A scientific viewing was consequently scheduled for October 26, 1967.

On Monday, October 23, 1967, Abbott informed newspapers in Victoria and Vancouver, British Columbia, of the film.[2] Furthermore, Jack Webster, a Vancouver radio station personality, interviewed Patterson and Gimlin on the air shortly after their arrival in Vancouver.[3] As news of the film spread, official word from the scientific community was eagerly awaited.

On October 26, 1967, both film rolls were shown at the University of British Columbia (UBC) to university scientists, B.C. Provincial Museum professionals, and other interested people. Casts of the footprints found at the film site were also provided for analysis. The UBC scientists were under strict orders not to comment one way or the other on the authenticity of the film. To this day, there has been no official comment from the university. Nevertheless, the sasquatch researchers in attendance recorded what transpired at the screening. The general "scientific" consensus was that the film was not sufficient evidence to prove the existence of sasquatch.[4]

Two museum professionals and a UBC spokesman provided statements to the press. The general tone of the statements was inconclusive or skeptical.[5]

A press screening and conference was held later that same day at the Hotel Georgia, Vancouver.[6] Attendance far exceeded expectations, and the press people demonstrated interest in the film. Subsequent newspaper reports were given high profile, but simply reflected the skeptical tone of the two museum professionals who provided statements after the UBC screening.

Because the "scientific community" was either skeptical (museum people) or noncommittal (UBC people), public interest in the new evidence quickly diminished. Hopes held by Green and Dahinden for government backing to conduct further research into the bigfoot phenomenon also rapidly faded.

A pillar at one of the entrances to the University of British Columbia beyond which the scientific fate of the Patterson/Gimlin rested on October 26, 1967.

Satellite photo of the University of British Columbia. It is a massive complex that is continually being expanded. Back in 1967 it would not have appeared quite like this, but still very large. Exactly what building was used to screen the Patterson/Gimlin film, I have not been able to determine. (Photo: Image from Google Earth. © 2008 Europa Technologies/Digital Globe.)

DETAILS AND ANALYSIS

1. The Cultivation of British Columbia
Through Green's and Dahinden's efforts, the Recreation and Conservation Minister of British Columbia, Kenneth Kiernan, had become interested in sasquatch research. Although highly skeptical on the issue, Kiernan apparently gave the evidence the benefit of the doubt. On September 19, 1967, Kiernan issued a public invitation for tangible evidence on the existence of sasquatch. Kiernan's involvement raised hopes that a government-sponsored expedition to find the sasquatch was a possibility.

2. Initial Newspaper Report
The first seed of doubt on the film's authenticity was planted by a British Columbia newspaper. On the day before the university screening, this paper stated Patterson had been rebuffed by university authorities in the United States. It was for this reason he was bringing the film to Canada. The statement was untrue. Certainly, people at Humboldt State College at Arcata, California had been contacted and they stated they were not interested. However, arrangements had already been made for the British Columbia screening. Furthermore, a representative from Humboldt State College, Professor W. J. Houck, did attend the UBC screening, although not as a college representative. His fare was provided by an Arcata, California radio station.

3. The Radio Interview
A part of this interview concentrated on why Gimlin had not shot the creature. The "pact," and the $100,000 reward were discussed. Jack Webster stated that, in the same situation, he thinks he would have shot the creature. Patterson replied, "I don't think you would have if you had seen the humanness of it. I think it would take a person with a little bit of murder in his heart to shoot something like this."

4. The Screening to Scientists
The people who attended this screening, to my knowledge, were: Dr. Ian McTaggart-Cowan, Dr. Beverly Green (both UBC scientists), Don Abbott, Frank Beebe, Charles Guiget (all professionals from the B.C. Provincial Museum), Roger Patterson, Bob Gimlin, John Green, René Dahinden, Bob Titmus (all sasquatch researchers). See the sidebar for qualification.

From information provided by René Dahinden and John Green, the following provides some insights on the proceedings.

Bob Titmus. (Photo, J. Green)

Attendance Qualification

After establishing the attendance shown in Point 4, John Green provided me with this list of attendees, which includes 34 people. However, it does not appear to show Dr. McTaggart-Cowan or Charles Guiget. I believe there was another UBC screening admitting additional people after McTaggart-Cowan and Guiget left, and this list was made at that screening. Furthermore, the list shows W.J. Houck, who I gathered was only at the press screening. It appears he attended both, as did many others.

Dr. Ian McTaggart-Cowan, the session leader, attempted to establish a physical height for the creature in the film. The purpose of the exercise was to determine if the creature was within the range of human proportions. McTaggart-Cowan used a yardstick to measure the creature's foot. He then determined that the height of the creature was 5.9 times the size of its foot. He then measured one of the plaster casts, arriving at 14.5 inches. Based on these two figures, the creature's height was 85.55 inches or 7 feet 1.55 inches. However, the doctor decided to reduce the size of the plaster cast to 14 inches. The reason for this action was to allow for foot slippage. Given the new figures, the creature height equaled 82.6 inches or 6 feet 10.5 inches. He then subtracted another 1.5 inches for an unknown reason to arrive at a final figure of 6 feet 9 inches.

Dr. McTaggart-Cowan also stated that the creature filmed was male because it walked like a human male. He was then questioned as to the creature's breasts. On this point he said that he could not see any breasts distinctly enough to identify them as mammary glands.

McTaggart-Cowan then made a profound statement:

"The more a thing deviates from the known, the better the proof of its existence must be."

With these few words, he tactfully told the audience that when it comes to bigfoot, the film is just not adequate proof.

In early 2004, I asked John Green to summarize the findings of the scientists at the screening and to provide his analysis. The following is his report.

Scientists among the group who watched the first public screening of the Patterson–Gimlin film in 1967 raised three negative comments about the creature on the film that continue to surface occasionally, despite the fact that two of them are totally wrong and the third is very questionable.

The first is that the creature has female breasts yet it walks like a man.

The second is that the creature has a sagittal crest, which is characteristic only of male gorillas.

The third is that no higher primate has breasts that are covered with hair.

In fact, it would have been an indication of a hoax if the creature did not walk "like a man." Among higher primates, only the human female has a walk different from that of the

John Green in his home office, 2003.

John Green's home in Harrison Hot Springs, B.C. I have visited him many times since meeting him in 1993. John is very precise in what he says and writes. He is the foremost bigfoot researcher in this field of study.

Behind the Scene

In all of the film frames, the creature has one or more of the following conditions: bent knees, bent body, nodded head. What we see is its "walking height." Between 8% and 8.5% needs to be added to walking height to arrive at standing height (actual height). While Dr. McTaggart-Cowan's calculation virtually defies reason and was hardly "scientific," if we add 8% to the height he established, we arrive at a standing height of 7 feet, 3.5 inches.[1] Remarkably, this is the **walking** height established by the North American Science Institution (Glickman). Whatever the case, acceptance of McTaggart-Cowan's figure (6 feet, 9 inches) still gives high credibility to the film. If the film was a hoax, a man of this height had to be found (reasonably difficult) and then enticed to act as the creature.

1. Here is the calculation: 6 feet, 9 inches equals 81 inches. 81 x 1.08 = 87.5 inches, which is 7 feet, 3.5 inches,

male, because wider pelvises are needed to give birth to human infants, which have exceptionally large heads.

As to the sagittal crest, its function is to provide an anchorage for large jaw muscles, and it is related to size, not to sex. Since the creature on the film is bigger than a male gorilla, lack of a sagittal crest would have been an indication of a hoax.

And female apes do have some hair on their breasts, even though they live only in hot climates. Apes adapted to climates as far north as the Bering Strait land bridge could surely be expected to have a great deal more.

None of the scientists at that first screening was a primatologist or physical anthropologist, and their opinions were asked for after a brief look at the film with no time for research or reflection, so their comments, while damaging, are probably excusable. There is no such excuse for their modern colleagues.

Here is what Dr. Grover Krantz, who was a physical anthropologist, had to say in his book *Bigfoot/Sasquatch Evidence:*

Human females generally walk rather differently from males, but there is no such contrast in apes. In our species, the female pelvis is relatively much wider at the level of the hip sockets than is the male pelvis. This results from the very large birth canal that is required for our large-headed newborns. Apes are born with much smaller brains, and their two sexes have more nearly the same pelvic design…

Of course the female sasquatch walks more like a man than a woman, and that is exactly how she should walk. (pp. 116–117)

…a sagittal crest is not a male characteristic…on the contrary it is a consequence of absolute size alone. As body size increases, brain size increases at a slower rate than does the jaw, so a discrepancy develops between these two structures. When jaw muscles become too large to find sufficient attachment on the side of the braincase, a sagittal crest develops.

That size threshold is regularly crossed by all male gorillas and a few females, by most male orangutans and by no other known primates. The evident size of the sasquatch easily puts their females well over that threshold, and

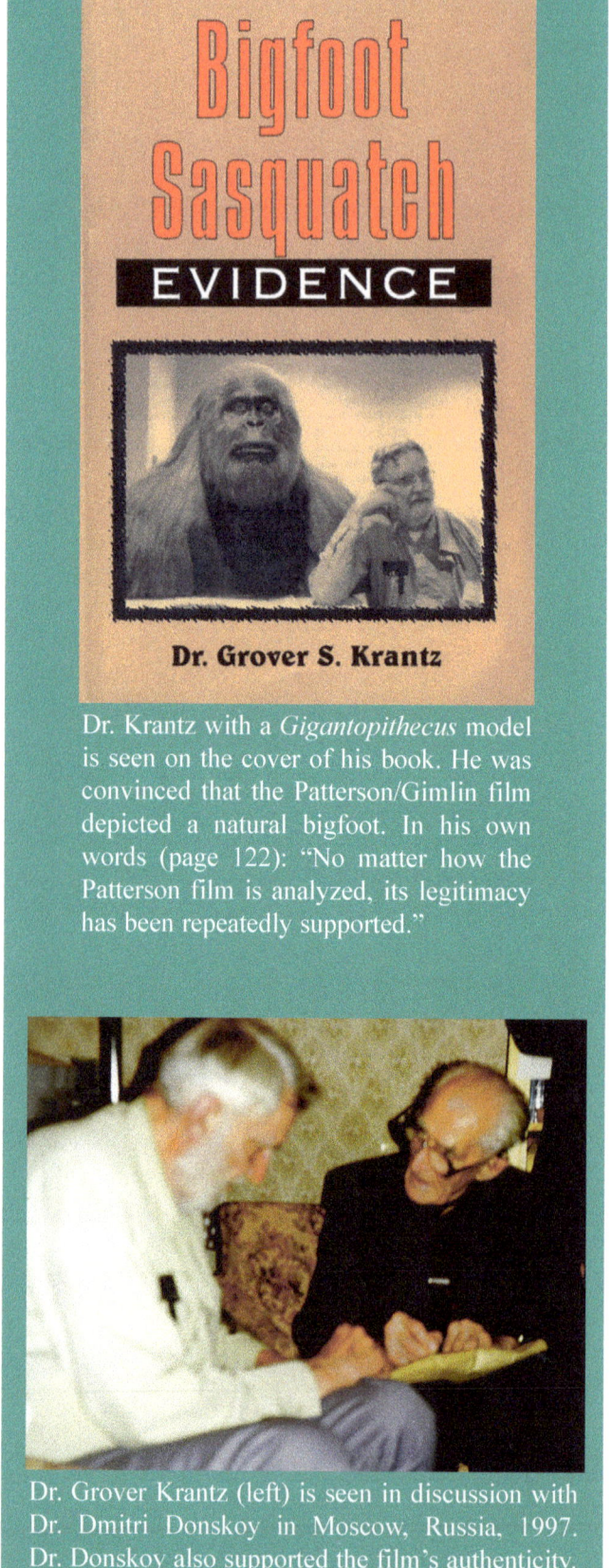

Dr. Krantz with a *Gigantopithecus* model is seen on the cover of his book. He was convinced that the Patterson/Gimlin film depicted a natural bigfoot. In his own words (page 122): "No matter how the Patterson film is analyzed, its legitimacy has been repeatedly supported."

Dr. Grover Krantz (left) is seen in discussion with Dr. Dmitri Donskoy in Moscow, Russia, 1997. Dr. Donskoy also supported the film's authenticity.

a sagittal crest would be an automatic development. (pp. 304–305)

That the species should have enlarged breasts at all (a human trait) is also a point of contention to some critics… But that they would be hair covered in a temperate climate seems perfectly reasonable to me. (p. 119)

5. Statements Made by Professionals to the Media
(Note: I show what the newspapers reported. I cannot vouch for the accuracy or intent.)

Don Abbott

I think most people assumed that it was a fake, and were content to let it go as that.

Further information is needed before the provincial government will aid in the search for the animal.

In the past, scientists have always laughed off reports of this kind, but the mere fact that a group of 50 or 60 scientists are prepared to get together and discuss it at least semi-seriously indicates a change in attitude. Before too long, someone is going to make a major attempt to get to the bottom on this thing somehow.

It is about as hard to believe the film is a fake as it is to admit that such a creature really lives. If there's a chance to follow-up scientifically, my curiosity is built to the point where I'd want to go along with it. Like most scientists, however, I'm not ready to put my reputation on the line until something concrete shows up—something like bones or a skull.

Note: There were not "50 or 60 scientists" at the UBC screenings of the film. This appears to be a misquote or something else was intended. by the statement.

Frank Beebe

I'm not convinced, but I think the film is genuine. And if I were out in the mountains and I saw a thing like this one, I wouldn't shoot it. I'd be too afraid of how human it would look under the fur. From a scientific standpoint, one of the hardest facts to go against is that there is no evidence anywhere in the Western Hemisphere of

The University of B.C.'s Museum of Anthropology is situated on the northwest side of the university endowment land. The building had not yet been constructed in 1967, but I would think the university had a museum of sorts somewhere. Last time I checked, there was absolutely nothing in the museum that mentioned the sasquatch, although I did not thoroughly study all of the totem poles. I did try to interest the museum people in a sasquatch exhibit, but to no avail. In all likelihood, some people associated with the museum were at one of the Patterson/Gimlin film screenings.

A couple of unique carvings that flank each side of the Museum of Anthropology. Native people had (and still have) great artistic talent. I often wonder why they have not carved (to my knowledge) a full size life-like sasquatch like these non-sasquatch carvings, rather than mythological renderings.

primate (ape, monkey) evolution—and the creature in the film is definitely a primate. So either a large primate got stranded in North America—or the film is a fake.

The animal's gait was too masculine, so was its build, yet it was purported to be a female. Also, it went around obstructions and not over them. Animals in the woods usually go over things, not around them.

The film gave the impression of great size in the creature—there was no accurate scale.

The background appeared quite genuine and there was no suggestion that the figure had been faked.

If it was fakery, it was very clever fakery, and it had to be a man in a monkey suit. We can't conclude that because there was nothing in the film to give us its proper proportions. If a series of footprints 1 ½ inches deep and 14 ½ inches long shown in a second piece of film by Patterson were, as he claimed, left by the creature, then only one conclusion could be drawn. If that is true and we have no means of proving or disproving this, then I've no doubt certainly what we saw was a Sasquatch.

In December 1967, Beebe had a bit of a change of heart and stated:

I'm not going to get into a newspaper hassle with Patterson. The Sasquatch is a phony and a fake as far as I'm concerned and that's it.

In April 1998, Beebe was contacted by bigfoot enthusiast Steven Harvey and asked for his opinion of the film when he viewed it in 1967. Beebe referred to the notes he had made on the day of the screening. He stated that he had written the following:

…a rather good, very interesting film. It just could be genuine and the darn thing for real, although the chance, indeed, the likelihood of a hoax is very high.

In early 2004, I talked by telephone to Beebe and asked for his current thoughts on the sasquatch, specifically with regard to sustenance. He was very positive and stated that he has absolutely no concerns as to the creature's ability to sustain itself. He pointed out that

Don Abbott is seen here on Blue Creek Mountain, California, in August 1967 "fixing" a track. When asked about his experience after he returned to Victoria, he said, "I was laughing at the whole idea on the way down, now I don't think it's a subject for mirth anymore. You realize that a scientist could ruin his reputation by going out on a limb and saying that the creatures exist. So I won't say I believe in them, but I am genuinely puzzled." (Photo, J. Green)

Arctic grizzly bear sustenance is infinitely less probable and the creature exists. This creature has only four months to find enough food to last an entire year. Beebe firmly concluded that sasquatch would be able to find more than adequate food sources all year round.

When I questioned him on the newspaper reports after the UBC screenings in 1967, he simply said, "paper does not refuse ink."

Beebe attended the opening ceremonies for my sasquatch exhibit at the Vancouver Museum, British Columbia, in June of 2004.

UBC Spokesman

It's either, a mighty big man in a monkey suit or it's a Sasquatch.

Dr. Ian McTaggart-Cowan

While no official opinion was expressed by the scientists at the first screening, during an interview in 1983, McTaggart-Cowan stated:

My memory is, there was nobody there that thought they were looking at a species of creature unknown to modern science.

6. The Press Screening and Conference

It is reasonably apparent that Dr. McTaggart-Cowan had determined that the film was a hoax before he had viewed it. Patterson had requested use of a room at the university for a press screening and conference. McTaggart-Cowan refused this request, apparently enforcing his policy of strict separation between science and show business. The session was therefore held at the Hotel Georgia in Vancouver. Press attendance was very good and high interest was demonstrated. There were many high-ranking newspaper people in attendance and some were quite impressed with the film. Also in attendance were Professor W.J. Houck (Humboldt State College, California) and David Hancock (zoologist/publisher). These two attendees made statements as follows:

Professor W.J. Houck

I'm not going to call it a hoax, yet the alternative is still to fantastic to accept. Where does that leave me? Darned if I know.

Houck said he doubted his college would back any attempt to launch a scientific inquiry, but the situation might vary according to the state of the

Three views of the Hotel Georgia and its prestigious Heritage Building Plaque. Patty was in good company, here is what is shown on the plaque:

CITY OF VANCOUVER

HERITAGE BUILDING

HOTEL GEORGIA

Architects: R.T. Garrow & John Graham, Sr.

This distinguished hotel was opened in 1927 by Edward, Prince of Wales, who later became King Edward VIII. Designed in the Georgian Revival style, for decades it was one of the main social centres at the heart of the city's downtown. Its celebrity guests included John Wayne, Nat King Cole, Elvis, and the Rolling Stones, and it was home to the Newsman's Club, and the ladies' Georgian Club and generations of UBC students in the downstairs George the Fifth pub. In 1998 it was restored by the owners Allied Holdings Ltd. with Bing Thom & Associates and Ray Littmann & Associates as architects.

college budget and the possibility of further evidence.

David Hancock

Dave Hancock wrote an extensive article on the film, and the sasquatch in general, that appeared in *Weekend Magazine* November 2, 1968 entitled, "The Sasquatch Returns." His summary of the film was as follows:

> If it was a hoax, then I join with most of the other observers in saying it was a very clever and sophisticated one.

Dave Hancock in 2004.

Until June 2004, publication of frames from the Patterson/Gimlin film was very limited. The first book that shows all of the clear frames and provides all of the scientific reports on the film under one cover is *Meet the Sasquatch*, published by Hancock House Publishers. I wrote this book in association with John Green and Thomas Steenburg, with major contributions from many main bigfoot researchers such as Yvon Leclerc, Dr. Jeffrey Meldrum, Dr. Henner Fahrenbach, Richard Noll, Daniel Perez, Doug Hajicek, Robert Morgan, and Joedy Cook. Dave Hancock was the driving force behind the publication of this work, and indeed suggested that the book accompany my sasquatch exhibit at the Vancouver Museum, June 2004 to February 2005. It should be noted that Hancock House is the primary publisher of sasquatch/bigfoot-related books.

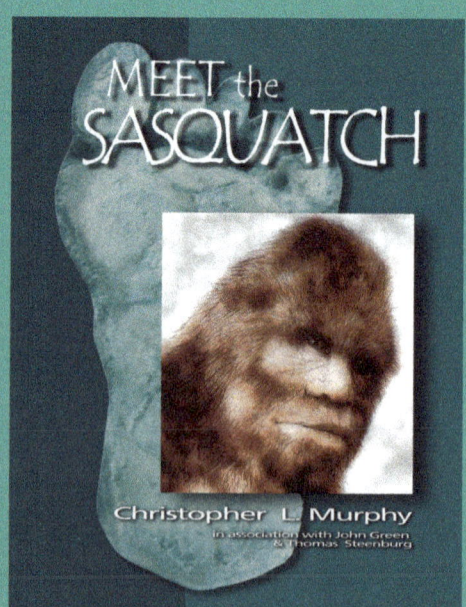

(Left to Right) Rick Groenheyde, your author, and Dave Hancock discussing *Meet the Sasquatch* in 2004. Rick designed the entire book (covers and contents). It's a beautiful book that will always be a tribute to Roger Patterson and Bob Gimlin.

8. The Canadian Controversy

In November 1967, Don Abbott and Frank Beebe provided official reports on the sasquatch to Kenneth Kiernan, the government minister. As none of the other scientists were permitted to say anything, Abbott and Beebe effectively had the stage to themselves. While their statements were not made on behalf of the provincial museum, they appeared to have the museum's sanction. Both reports fell short of recommending a government-sponsored expedition to find the sasquatch.[1]

John Green and René Dahinden attempted to turn the tide by appealing to Dr. Clifford Carl, curator of the provincial museum. In February 1968, a slow-motion version of the film showing the creature was professionally prepared with enlargements and stop-frame action. This film version was presented to Dr. Carl along with about 100 members of the Victoria Natural History Society and a group of University of Victoria students. After the viewing, Dr. Carl said that the enhancements added a lot to the film. He then remarked, "But they haven't done much to clear-up some of the doubts in the public's mind." Asked as to his professional opinion on the film, Dr. Carl replied, "I'm still sitting of the fence." Later, Dr. Carl's *official* museum statement on the whole controversy did nothing to change the general situation. He merely stated, "The Provincial Museum feels it is impossible to determine with any degree of certainty either that the Sasquatch is real or that it is a hoax."[2]

The controversy and uncertainty over the Patterson/Gimlin film had a *negative* influence on Kenneth Kiernan. This government minister had taken the first steps toward a government-sponsored search for the creature.[3] Now, all of the groundwork undertaken by Green and Dahinden was rapidly eroding. Ironically, the most convincing evidence of the creature's existence was working **against** government involvement.

In early 1969, the provincial museum issued a statement that it was not actively investigating the sasquatch issue and that it had no intention of sending an expedition to look for sasquatches on the basis of the evidence presently available.

Considering the implications of confirming the reality of sasquatch, the Canadian controversy ended very quickly. Press coverage ran its course, and the sasquatch soon became "yesterday's news." It certainly appears, however, that the bigfoot issue had lingered in the minds of both Frank Beebe and Don Abbott. Dr. John Napier, in his book *Bigfoot* (1972), credits Beebe and Abbott with coming up with "a most ingenious model for Sasquatch feeding habits."[4]

Dr. Clifford Carl

For the Record

In 1970, *National Wildlife* magazine managing editor George H. Harrison subjected Roger Patterson to a polygraph test (lie detector). Here is what Harrison wrote in his article "On the Trail of Bigfoot" that was featured in the October–November 1970 issue of the magazine:

"Before printing the [Patterson/Gimlin] story a *National Wildlife* editor flew to the West Coast to interview Patterson, who believed so strongly in Bigfoot and the photographs he had made that he instantly agreed to take a lie detector test. The results convinced the experienced polygraph operator that Patterson was not lying."

I know, of course, that polygraph tests are not foolproof. However, I think *National Wildlife* would have made certain the test was done properly as their reputation was at stake. I don't have the same confidence in a recent test concerning Robert Heironimus who says he played the part of the creature in the film and, we are told, passed a lie detector test.

The complete *National Wildlife* article link:

http://forum.hancockhouse.com/media/natwildlifefinaluse.pdf

DETAILS AND ANALYSIS

1. The Official Reports

Don Abbott's Official Report

Abbott, who had seen actual footprints firsthand, maintained a position of "soft skepticism." Unfortunately, Abbott's stand was a classic case of "too little, too late." Nevertheless, Abbott did support (as opposed to recommend) further investigation.

As we can see, Abbott's statement is one of total

> **DONALD N. ABBOTT**
>
> **REPORT TO THE MINISTER ON RECENT INVESTIGATIONS INTO THE SUPPOSED "SASQUATCH"**
>
> **BRITISH COLUMBIA PROVINCIAL MUSEUM, VICTORIA**
>
> **NOVEMBER 9, 1967**
>
> I have now come to feel that this is a problem which should finally have serious investigation and that it deserves to have some money spent on it even if nothing more should be accomplished than uncovering a hoax. The decision whether or not Provincial Government money ought to be spent on such an investigation is obviously one that can be made only at Cabinet level. My personal feeling is that if nobody else is going to follow this further, then probably we should.
>
> Under the present circumstances, if it is decided that the Province should continue to investigate the sasquatch phenomena at all, there are a number of possible courses open. None of them can be guaranteed either to produce a sasquatch or to prove their non-existence.
>
> First of all, more information could be collected and collated from members of the public who claim to have seen sasquatches or have other kinds of evidence apparently relating to them. This could be accomplished by more public appeals and requesting information of licensed game guides and outfitters and of Provincial Conservation Officers and checkpoint staffs. If this course of action is followed, it would require the appointment of someone who could spend full time doing this work and be free to travel in order to check reports and interview people.

Who Was Don Abbott?

Don Abbott passed away in the summer of 2005. The Victoria *Times-Colonist* published his obituary on July 31, 2005. It reads, in part:

> Abbott, Donald Neil. Died peacefully on Thursday, July 28. He was born April 30, 1935. Don is survived by his wife Maria, children James, Ian and Sarah; daughter-in-law Jo-Ann; granddaughters Railey, Brynn, Chloe and Augusta; nephews Craig and David; niece Lynn and their children. Don was raised in Vancouver where he attended King George High School and the University of British Columbia, followed by postgraduate work at the Institute of Archaeology, London University and Washington State University, Pullman. Don was loyal to all his friends and a faithfully committed husband and father. He was fond of his Scottish heritage, loved wearing his kilt at formal occasions, a member of Saltire Society Victoria and for many years of the Victoria Branch of the Scottish Country Dance Society. He enjoyed photography and scuba diving, especially with the Underwater Archaeological Society in their earliest years. In 1960 Don joined the staff of the Royal BC Museum as the first professional archaeologist in BC. In his 35-year career, Don continuously and selflessly promoted the discipline of archaeology. He conducted excavations on sites on southern Vancouver Island that contributed significantly to an understanding of Coast Salish history. Under his direction, the Museum became the provincial centre for the documentation of archaeological sites and the storage of artifacts and associated data, held in trust for First Nations. He was a member of the Archaeological Sites Advisory Board, of the committee that created Ksan Cultural Centre in Hazelton, and worked on the Exxon Valdes Recovery Programme. Don loved words and language. As writer and editor, he enhanced the stature and scientific rigor of publications produced under Museum imprint. His contributions remain important references in the literature, notably "The World is as Sharp as a Knife ," an anthology memorializing his friend, the late Wilson Duff. Don brought high moral principles to his job. During his time, he was the primary conscience of the Museum, establishing and defending humanistic, scientific and literary standards. A modest man, he felt no need to proclaim his achievements, but many colleagues remain honoured to have learned from and with him and continue to remember his selfless commitment to the unique cultural history of British Columbia. Don was a man of great dignity, humour and intelligence.

(As far as I know, Abbott never got involved further with the sasquatch issue beyond what I show in this work.)

Certainly, neither I nor any of the other members of the museum staff could add these responsibilities to our present duties. Mr. John Green has on a number of occasions expressed a desire to be contracted or appointed for this work himself. Upon consideration it seems to me that if such a program is decided upon he probably would, in fact, be a most appropriate choice. From a collection of reported information like this, we could hope to construct (given the creature's existence) a mosaic picture of its appearance, habits and distribution.

Of course, all of the members of the sasquatch hunting fraternity (Green, Titmus, Dahinden, Patterson, Gimlin, possibly others) have been anxious to receive financial support for their activities and are eager to be taken on any professional expedition, preferably for a fee. It is quite obvious that this is a matter which must be treated with a great deal of caution. On the one hand, they can rightly point out that they all have spent a great deal of their own time and money on this obsession and have thereby acquired a considerable amount of information of greater of lesser value. To some degree, at least any professional research into the subject would have to begin with the information they have collected, and of course we would owe credit to them, especially to John Green, for their persistence which has finally given the impetus to our investigation. On the other hand, it is hard to see how in fact they are actually owed anything, let alone the degree of glory and wealth about which some of them seem to be daydreaming.

Frank Beebe (b. 1914) now retired, is seen here in a recent photo. During his working career, he was a zoo director and then a scientific illustrator and museum illustrator. This last position he held at the British Columbia Provincial Museum (later the Royal Museum) at the time the Patterson/Gimlin film was presented.

Beebe is a highly accomplished and acclaimed artist. His hobby is falconry and he is noted as the quintessential falconer. His many remarkable paintings show birds of prey. I have seen a number of these and am truly amazed at the astounding reality of the birds depicted. He has written several books and numerous articles on falconry.

indecision. He does not make a firm recommendation. He does, however, suggest a course of action, actually naming John Green to run a program. He then throws in a red herring by implying possible ulterior motives on the part of researchers in the "sasquatch hunting fraternity." It would indeed be interesting to know what went on behind closed doors at the B.C. Provincial Museum.

Frank Beebe's Official Report

Beebe addressed the improbability of sasquatch creatures based on what is known (or unknown) about the creature. His report implies that the creature probably does not exist and therefore the government should not get involved in any action to prove or disprove its existence. The conclusion to his official report to Kenneth Kiernan is as follows (I do not have the first part).

This is the main entrance to the new Royal B.C. Provincial Museum. The word "Royal" was added upon a visit by Queen Elizabeth II in 1986. The first museum building opened its doors in 1886. The building seen here was completed in 1967. A glass lobby addition was completed in 1997. (Photo, Ryan Bushby)

FRANK L. BEEBE PARTIAL OFFICIAL REPORT

…Thus, while the existence of the mythical "sasquatch" remains extremely improbable, it is not entirely impossible. The anthropologists who have looked at the tracks tell us that, if made by a real animal, such a creature must be, not merely a primate, but a hominid primate, and thus very closely related to the human. As a final critique, it must be pointed out that, if such a creature exists, it must differ very profoundly from man, and from most other primates as well in perhaps the most significant way of all. To be credible at all, the sasquatch must be incredible. It must be, for a primate, incredibly stupid and must have an intelligence scarcely on a level with that of a deer or a goat, and not as high as that of a bear. It builds no shelters, stores no food, makes no fire, has developed no culture and does not even have sufficient wits to learn how to peel bark to make use of the best natural foods produced in the region in which it lives.

2. Dr. Clifford Carl's Reaction

The issue was with the Patterson/Gimlin film, *not* the sasquatch creature in general. Carl appears to have considered this difference and chose to stay on the safe side. Had his official statement referenced the "Sasquatch film," rather than just "the Sasquatch," the film would have received a degree of uncertain credibility. However, this condition would have undermined Beebe's statements. It is apparent Carl did not wish to do this.

3. The Influence on Kenneth Kiernan

Kiernan was in a difficult position. He had been positively influenced by footprints and sighting reports. He was now being negatively influenced by actual movie footage. If he went forward with any plans for government involvement, this action would imply acceptance of the film. With the public heavily influenced by Frank Beebe and Don Abbott's statements, Kiernan's logical alternative was to pull out of the issue. Defending the use of government funds to support sasquatch research would have been difficult in the first place. Now, with a possible hoax being provided as evidence, such a defense would be impossible.

4. Napier's Book

Napier stated the following in his book (p. 172):

> Frank L. Beebe and Don Abbott of the Provincial Museum, Victoria, B.C., have come up

Kiernan's Last Word

On November 5, 1967, John Green wrote to Kenneth Kiernan confirming his belief in the authenticity of the Patterson/Gimlin film and emphasizing that:

"the creature fits readily in every detail within the pattern of eye-witness descriptions."

Green asked the minister for a meeting to further discuss the sasquatch issue.

Kiernan replied on November 17, 1969 and his bottom line was:

"I trust you appreciate that we operate on fixed budgets and we are not in any position to spend any further sums of money, even small sums, on this project at the present time."

He then mentioned a glimmer of hope that never materialized:

"We will continue to collect any pertinent data that becomes available and may be able to develop a further line of investigation that might prove useful."

Aside: I will mention that there is other information in Kiernan's letter, but it is so idiotic I am embarrassed to present it here because he was a minister in the Canadian province in which I reside. CLM with a most ingenious "model" for the Sasquatch's feeding habits. They base it on the life-style of the wolverine, a weasel-like mustelid of large proportions (3 ft. or more including tail). This extraordinarily interesting animal has a wide home-range of 200 miles or more, is largely carnivorous and rapacious with it (not for nothing is it known as the 'glutton'). Wolverines, broadly speaking, occupy the same habitat as the Sasquatch. A particularly relevant aspect of their behaviour is that wolverines cache their food in natural 'deep-freeze' lockers above the snow line, winter and summer. They range upwards into the snowfields when food is scarce at lower altitudes, open their lockers and (presumably) carry the food down to altitudes where it gradually thaws. This model might explain how the Sasquatch survives through the winter, and why so-called Sasquatch footprints have been observed at high altitudes by skiers and snow-mobilers; equally of course, the habits of the wolverine might account for the very existence of these amorphous tracks. Let me give full rein to imaginative speculation: could deep-freeze behaviour patterns also explain the apparently inexplicable occurrence of Yeti footprints high above the snowline in the Himalayas?"

WALKING ON EGGS

When we read the responses of anthropologists and related professionals on the sasquatch/bigfoot issue, we become a little disheartened because of the "arms length" or negative approach they take. Few have taken the trouble to have a really good look at the evidence. Indeed, the vast majority completely ignore any issue that is not totally "black and white."

John Green provided an excuse for the professionals at the first screening of the Patterson/Gimlin film. René Dahinden was far less charitable—the words he used to describe professionals are not in your spell-checker.

Nevertheless, I do believe these scientists would really love to jump in and have a look at the sasquatch issue. I even muse that some of them have bought books about the sasquatch—carefully making sure the book is put in a brown paper bag.

The problem with the field of anthropology is that anthropologists don't have a wide selection for employment. Most of these professionals are employed either directly or indirectly by governments. Universities top the list, and there are only a limited number of positions available.

However, because the field is highly interesting, there are many candidates for positions, far more than the number of positions available. In other words, "supply greatly exceeds demand."

As a result, most anthropologists "walk on eggs." If they are seen dabbling in anything "suspect," such as the sasquatch, then they run the risk of ruining their reputation and losing their position, or at the least being passed over for advancement.

In my opinion, the few employed anthropologists who have taken on the sasquatch challenge are truly intrepid individuals. We should really not expect to get anthropologists involved until after they retire and are no longer "at risk," as it were.

While I do hold hope that we will be able to turn the tables in the sasquatch issue with "more of the same" (footprints and sightings), this approach is getting very thin. We definitely need something more for the scientists. Here, I am referring to actual hard evidence, not photographs. If the Patterson/Gimlin film were taken in recent years, its credibility would be greatly diminished (or totally discounted) because of the increased difficulty to detect fabrications.

9. Patterson and Gimlin in Hollywood, New York & Atlanta

In early November 1967, Patterson, Gimlin, and DeAtley took the film to Universal Studios, Hollywood, California, to have it reviewed by special effects people. Technicians were asked if they could reproduce (duplicate) a creature such as that seen in the film. Their response, although marginally qualified, was negative.[1]

Later that month, Patterson was contacted by *Life* magazine regarding a possible photo story on the film. Before committing to a contract, however, the magazine people wanted their own scientific appraisal on the film's contents. To this end, *Life* arranged for a scientific screening of the film at the American Museum of Natural History in New York City. Patterson was invited to present the film at the museum.

Patterson immediately seized the opportunity and went to New York, accompanied by Gimlin and DeAtley. At the screening, it became evident that the scientists were just doing a favor for the magazine. They were less than enthusiastic at the viewing. It appeared as though they had made up their minds that the film was a hoax long before Patterson had arrived. The scientists viewed the film only once, made no measurements, and did not ask for any stop-framing. Patterson, Gimlin and DeAtley were asked to wait outside after the viewing. Within fifteen minutes the scientist announced, "It [the film] is not kosher because it is impossible."[2] With this news, the magazine people started to lose interest in the project.

However, they decided to get a second opinion from animal specialists at the Bronx Zoo. The people at the zoo were somewhat more open-minded. They requested at least two replays and asked for several stop-frames. Their final opinion, however, was negative. They reported that there was "something wrong" with the film. They did not substantiate their opinion with any additional information.[3] With two rejections, as it were, *Life* magazine lost interest and did not enter into a contract with Patterson and his associates.[4]

While in New York, Patterson contacted Ivan T. Sanderson,[5] a noted zoologist who had long been involved in the search for America's abominable snowman. After seeing the film, Sanderson got in touch with *Argosy* magazine people who had expressed interest in the new evidence. Working with Sanderson, *Argosy* assembled four noted scientists, together with the Director of Management Operations for the U.S. Department of Interior, to see the film. Also present were journalists and newspaper people, including a member of the

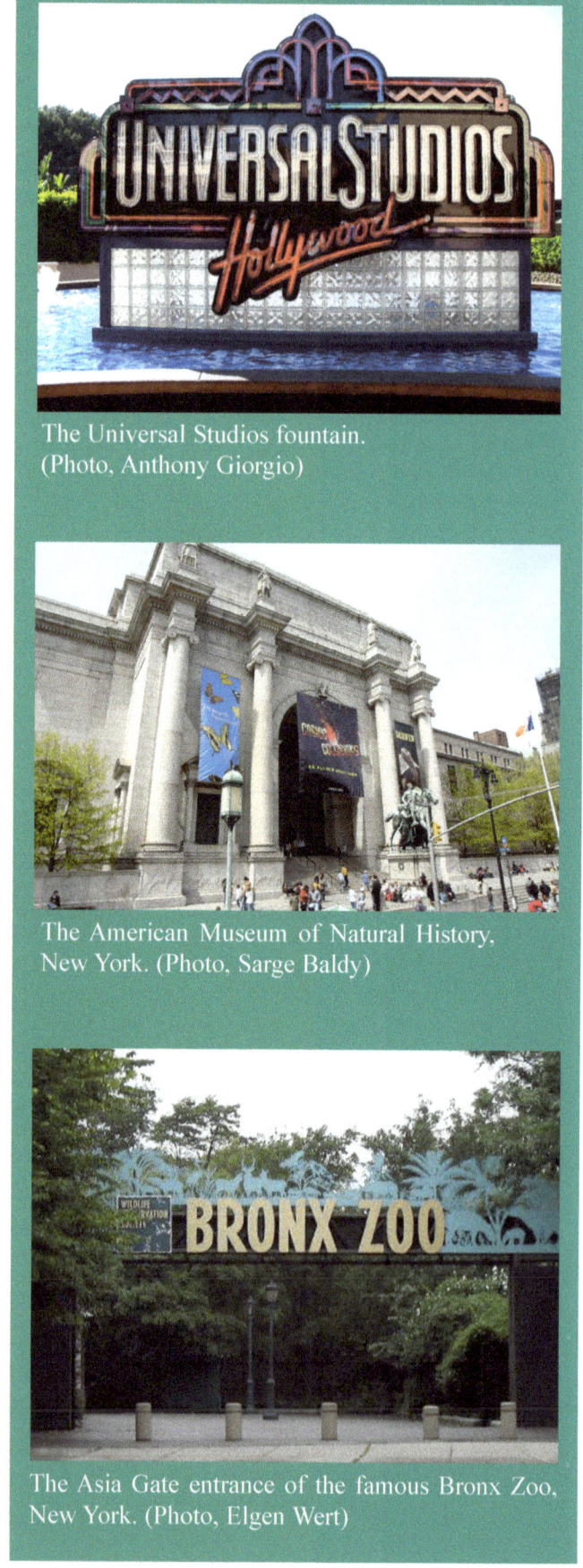

The Universal Studios fountain. (Photo, Anthony Giorgio)

The American Museum of Natural History, New York. (Photo, Sarge Baldy)

The Asia Gate entrance of the famous Bronx Zoo, New York. (Photo, Elgen Wert)

editorial staff of the National Geographic Society.

The film itself, and still photographs from the footage were shown and examined under high magnification. The only notable outcome of the meeting was that no one said he thought the film was a hoax.[6] The *New York Times* people were apparently unimpressed. There was no coverage of the event in this paper. The November 16, 1967 edition of *The Post* had an article on page 12. The National Geographic Society also completely dismissed the evidence.[7]

The group then travelled to Atlanta, Georgia where another screening was held for Dr. Osman Hill, head of the Yerkes Regional Primate Research Center at Emory University. Dr. Hill stated that if the bigfoot film was a hoax, it was extremely well done and effective. He also stated that the evidence was strong enough to mount an expedition to search for further evidence.[8]

The *Argosy* people remained interested in the film, so Sanderson wrote an article about it for this magazine. The article was featured in the February 1968 issue with cover-story prominence.[9]

Later in 1968, the film was shown to scientists at the Smithsonian Institution. It is my understanding that Dr. John Napier arranged this for the Smithsonian's General Conference, and there were about twenty scientists present. Although there was no official statement made by the Smithsonian on the film, it has been learned that the opinions expressed were highly negative, save that of Dr. Napier.[10]

DETAILS AND ANALYSIS

1. Universal Studios
The group met with Dale Sheets, head of the Documentary Film Department, along with top technicians in the Special Effects Department. After studying the film, their exact response on the question of reproducing (duplicating) the creature in the film was as follows:

> "No…maybe if you allotted a couple of million bucks, we could try, but we'd have to invent a whole set of new artificial muscle, get a gorilla's skin and train an actor to walk like that. It might be done, but offhand, we'd say it would be nearly impossible."

In 1969 John Green took the film to Disney Studios. He was informed that the Disney people had already studied the film. It is not known for certain if this occurrence is connected with Patterson's previous

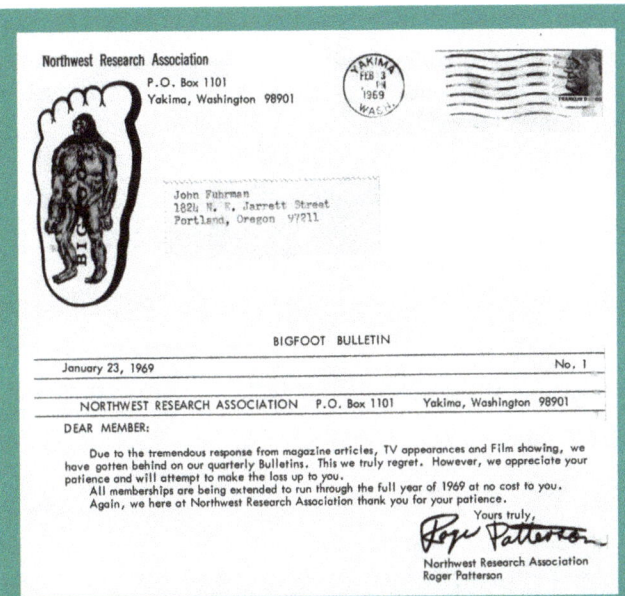

Top portion of a Bigfoot Bulletin issued by Roger Patterson under the name of his company, the Northwest Research Association. I don't know how many he issued, but I have found a few. In this issue (January 23, 1969) he apologizes for getting behind in providing bulletins, and informs all members of his association that he is extending their memberships to run through the full year of 1969. There is no doubt in my mind that Roger tried to meet his various commitments, but often bit off more than he could chew.

The Smithsonian Institution in Washington, D.C. It was created with funds provided in the will of James Smithson, an Englishman, who died in 1829. Smithson's provision was for an institution, "at Washington, under the name Smithsonian Institution, an establishment for the increase and diffusion of knowledge among men."

visit, although it probably was. Disney Studios informed Green that if they wished to create a creature of this nature they would draw it. Green was then referred to Janos Prohaska, a top Hollywood ape imitator. Prohaska stated that the creature was not a man in a conventional gorilla-type suit because muscle movement can be seen. Any suit would have to be skintight to show this movement.

The fact that Patterson took the film to Hollywood is noteworthy. If he knew the film was a hoax, it is unlikely he would have taken this action.

It also needs to be mentioned here that the film was again taken to Disney Studios for examination by Peter Byrne in 1973. The chief technician reported, "If it is a fake, then it is a masterpiece and as far as we are concerned the only place in the world where a simulation of that quality could be created would be here, at Disney Studios, and this footage was not made here." (Peter Byrne, 1975. *Bigfoot — Monster, Myth or Man?* p. 115.) Furthermore, in 1983 Bruce Bonney, who worked with René Dahinden, took the film to Hollywood and showed it to special effects people. Bonney does not wish to have the results of his meeting publicized, however, no firm conclusion was reached on the "nature" of the creature.

2. The Museum Viewing
Few people would be impressed with a simple viewing of the Patterson/Gimlin film. Virtually all that can be seen is what appears to be a hairy/furry person walking rapidly along a sandbar. As there is nothing on which to base the size of the creature, one instinctively gives it normal human being proportions. Because we have been conditioned to think of the creature as being very large, one's immediate reaction is disappointment. The museum scientists, which included Dr. Richard Van Gelder and Dr. Harry Shapiro, could have been, and should have been, much more "scientific."

3. The Bronx Zoo Viewing
The Bronx Zoo people did give the film a chance. Their opinion that there was "something wrong" with the film, however, is a very poor response. As they did not say what was wrong, they evidently did not know what was wrong.

4. *Life* Magazine
In December 1972 (over five years later), *Life* magazine had a change of heart. One of their reporters spent several days with René Dahinden preparing a story on the film. The story was to be a six-page color spread. Unfortunately, the magazine ceased operations before the story went to print.

Test of Time

Every attempt to duplicate the Patterson/Gimlin film to date has been a dismal failure. The results were not only ludicrous, they were laughable. On can tell immediately that the creature is just some clumsy actor in a fake fur costume.

A Typical Reaction

I think René Dahinden's first reaction to the film is very typical and provides an insight into what others thought. Here is what he said:

"I knew what I was going to see, I'd had the thing described often enough, but it still gave me a hell of a shock when I saw it. Your first reaction is, 'Ah comeon…' you know, looking for the zipper in the fur suit. But then you start looking at it one thing at a time…"
(D. Hunter with R. Dahinden. *Sasquatch/Bigfoot,* p. 116.)

I must admit that I felt the same when when René first showed me the actual film. However, I took his word for it that there was more than meets the eye. Unfortunately, the professionals were not prepared to deal with this sort of thing, so they simply didn't deal with it.

Ivan T. Sanderson

Looking at things from a "scientific" standpoint, Patterson's immediate association with Ivan T. Sanderson, was probably not the smartest thing to do. Sanderson was renown for being somewhat "far out." A private letter that has been floating around for about 40 years, said to be written by someone "in the know" in Smithsonian circles, stated the following about him:

"Sanderson doesn't have a high reputation as a biologist; of course he doesn't exactly claim it for himself—he calls himself an ecologist, and I think he specializes in plants, certainly not primates. Also he has an oversize evolutionary ax to grind, and would rejoice to find proof of some living subhuman creature of the missing-link variety, which he says will absolutely disprove the Bible."

5. Ivan T. Sanderson

Sanderson had corresponded with Patterson for about six years prior to their meeting in New York. Patterson states in his book, *Do Abominable Snowmen of America Really Exist?*, that he became interested in bigfoot after reading an article by Sanderson. The article, entitled *The Strange Story of America's Abominable Snowman*, appeared in the December 1959 edition of *True Magazine*. In 1961, Sanderson authored a book entitled, *Abominable Snowmen — Legend Come to Life*. Sanderson stated in his *Argosy* article that he had been researching information for this book since 1930.

6. The Meeting Arranged by Sanderson and Argosy

The scientists at this meeting besides Sanderson were (1) Dr. A. Joseph Wraight (human ecologist), Chief Geographer, U.S. Coast and Geodetic Survey; (2) Dr. John R. Napier (primate biologist), Director, Primate Biology Program, The Smithsonian Institution; (3) Dr. Vladimir Markotic (physical anthropologist), Associate Professor of Archeology, University of Calgary, Alberta, Canada; (4) Dr. Allan Bryan (anthropologist), Professor of Anthropology, University of Alberta, Edmonton, Alberta, Canada. Others present of note were: Mr. N.O. Wood, Jr., Director of Management Operations for the U.S. Department of the Interior (representing Stewart Udall, Secretary of that department), and Tom Allen, writer and editor on the editorial staff of the National Geographic Society. Official statements from two of these professionals are as follows.

Dr. A. Joseph Wraight

The presence of large, hairy human-like creatures in North and Central America, often referred to as Sasquatch, appears very logical when the physiographic history of the northern part of this continent is considered. The statement often made that monkey-like creatures were never developed in North America may easily be discounted, for these creatures are more humanlike than ape-like and they apparently migrated here, rather than representing the product of indigenous evolution. The recent physiographic history of the polar edges of North America reveals that the land migration of these creatures from Asia to America is a distinct and logical possibility.

The compelling reason for this distinct possibility is that a land bridge between Asia and North America is known to have existed several times within the last million years, at various intervals during the Pleistocene or Ice Age.

Excerpt of a letter from Dr. A. Joseph Wraight to Roger Patterson (shown in Patterson's Bigfoot Bulletin, January 13, 1969), reprinted below.

U.S. DEPARTMENT OF COMMERCE
Environmental Science Services Administration

Dear Mr. Patterson:

I would like to take this opportunity to congratulate you on the fine work you are doing toward the solving of the "Bigfoot" scientific mystery. The question as to whether or not these subhuman creatures do or do not exist in the rugged western highlands of North America has plagued science for too many decades, and it is heartening to see that you are doing something about the matter. The act of solving this riddle requires a great deal of patience, courage and effort, and possibly for this reason many scientists have long avoided or ignored the question. Apparently you and your organization have the fortitude it takes to track down and capture one of these creatures, so that direct observation could be made by man, and the mystery could be cleared up once and for all.

To me, of course, there is no real mystery. As you know, I believe firmly in the existence of these creatures, and I would like to see you capture one of them so the remainder of the scientific community could observe it and become also staunch believers. Then I believe that a great step would be made in scientific knowledge. Certainly the uncontested existence of another type of subhuman creature would be recorded, and the exact nature of his biological structure and functions would be known. Perhaps the entire gamut of evolutionary concepts might undergo alteration on account of this, and geographical concepts of species migrations and changing environmental conditions of the past might have to be revised. Accordingly, I strongly urge you to keep up your good work in searching for the truth in this matter, for such truth is the ultimate aim of all science.

Sincerely,
A. Joseph Wraight
Chief Geographer, C&GS

The land bridges, both on the north and south sides of the Bering Sea, were admirably suitable for migration several times during the Ice Age.

It appears then, that these hairy, human-like creatures, sometimes called Sasquatch, could easily have migrated to North America at several times during the Ice Age. This is particularly plausible when it is considered that conditions were mild in the area when the land bridges existed. These creatures could have then found conditions along the way similar to their Asian mountain habitat and could naturally have migrated across the bridges.

Dr. John R. Napier (d. 1987)

First, I observed nothing that, on scientific grounds, would point conclusively to a hoax.

I am satisfied that the walk of the creature shown in the film was consistent with the bipedal striding gait of man (except in the action of the feet, which were not visible). I have two reservations which are both subjective: First, the slow cadence of the walk and the fluidity of the bodily movements, particularly the arms, struck me as exaggerated—almost self-conscious in comparison with modern man; second, my impression was that the subject was male, in spite of the contrary evidence of heavy, pendulous breasts.

The bodily proportions of the creature, as far as could be seen, appeared to be within normal limits for man. The appearance of the high crest on top of the skull is unknown in man, but given a creature as heavily built as the subject, such a biomechanical adaptation to an exclusively fibrous raw vegetable diet is not impossible. The presence of this crest, which occurs only in male non-human primates, such as the gorilla and the orangutan, tends to strengthen my belief that the creature is male. Finally, it might be supposed that a creature with a heavy head, heavy jaw and musculature and a massive upper body would have a center of gravity placed at a higher level than in man. The position of the center of gravity modifies the gait and the easy stride shown in the film is not in harmony with a high center of gravity.

Some of the questions I have raised might be solved by a scientific frame-by-frame analysis of the gait and body proportions, and a study of the joint angulation and limb displacements. This should be done.

The opinions I have expressed on this remarkable film are those of an expert witness rather than a member of the jury."

Napier states the following in his book, *Bigfoot* (1972).

I am convinced that the Sasquatch exists, but whether it is all that it is cracked up to be is another matter

Excerpt of a letter from Dr. John R. Napier to Roger Patterson (shown in Patterson's *Bigfoot Bulletin*, January 13, 1969), reprinted below.

SMITHSONIAN INSTITUTION
United States National Museum
Washington D.C. 20560

Dear Roger:

As you know, I have had a scientific interest in Bigfoot problems for a considerable time and I have always considered the Bigfoot the most interesting of all the hairy men because, unlike most of the others, Bigfoot is cast in a human image. The footprints for instance, of which I have seen a number, are humanoid and as far as I can see without having had the opportunity of studying them closely or for any length of time, they are consistent with the human type of walking gait in soft soil (which differs from the gait on resistant surfaces).

To me, the factors which makes it a possibility are (1) the likelihood that early humans could have entered America via the Bering Strait land bridge on a number of occasions during the last half million years. There was considerable two-way traffic among mammals and there is no earthly reason why man shouldn't have been among them; after all, they would merely be following their food supply! (2) Man is supposed only to have reached America in the last 10-15,000 years; it has always been a matter of surprise to me that he waited so long?

Sincerely

John R. Napier, Director
Primate Biology Program

altogether. There must be something in north-west America that needs explaining, and that something leaves man-like footprints. The evidence I have adduced in favor of the reality of the Sasquatch is not hard evidence; few physicists, biologists or chemists would accept it, but nevertheless it is evidence and cannot be ignored.

7. National Geographic
This organization finally got around to looking at the sasquatch controversy in January 2005 by contracting a television program producer to produce a documentary. Unfortunately, they allowed the producer to take a light-hearted approach to the phenomenon. In other words, they made a laughing stock of the evidence, those who have witnessed the creature, and the many dedicated people who have studied the evidence.

8. Dr. Osman Hill's Statement
Dr. Hill apparently took the time to thoroughly study the creature in the film, and his conclusion (right or wrong) echoed those of previous Hollywood film experts. The following is his official statement on the film:

> The creature portrayed is a primate and clearly hominid rather than pongid. Its erect attitude in locomotion, the gait, stride and manner of that locomotion, as well as the relative proportions of pelvic to pectoral limb, are all manifestly human, together with the great development of the mammary glands. This does not, of course, preclude the possibility that it is indeed a *homo sapiens* masquerading as a hairy "giant."
>
> All I can say, at this stage, is that if this was a masquerade, it was extremely well done and effective.
>
> Without tangible evidence in the form of skeletal parts, a cast of the dentition or similar physical material, I cannot pronounce beyond this group. However, the most interesting evidence they have so painstakingly produced should serve to stimulate the formation of a truly scientific expedition to the area, with the object of obtaining the required physical data.

9. The *Argosy* Magazine Article
This was the first major magazine article written about the film. It contained a number of different frames from the film, and a quoted interview with Patterson and Gimlin. There are inconsistencies in the article between what was known at the time and information later revealed through investigations. The cover of the magazine shows frames from the film along with a photograph of Patterson and Gimlin on horseback. Gimlin is dressed-up like a 19th century Indian guide. He is wearing a long black wig with a white headband.

I originally thought the photograph was a publicity stunt by *Argosy* magazine, and had some trouble justi-

Bob Gimlin (left) and Roger Patterson in the photograph used on the cover of *Argosy* magazine, February 1968.

Note on Dr. Markotic

Although Dr. Vladimir Markotic did not make a statement at the time he first saw the film, he and Dr. Grover Krantz edited a book entitled *The Sasquatch and Other Unknown Hominoids* (Western Publishers, Calgary, Canada) that was released in 1984. I think the book itself is testimony that he believed the issue deserved scientific attention. I will mention, however, that Thomas Steenburg (who was good friends with Vladimir in Alberta) told me he did believe in the existence of the creature.

fying how Gimlin would allow himself to be dressed-up that way in connection with the film. Partly of First Nations heritage, Gimlin dressed this way for the local Appaloosa shows in which he frequently participated. Patterson wanted a photograph of this nature for effect in his planned film documentary. It was taken by a friend about one year before the Bluff Creek event at a spot about ten miles from Patterson's home. It was subsequently provided to *Argosy* for the magazine article.

10. The Smithsonian Institution Viewing

The following is a summary of what is believed were the opinions and thoughts of the Smithsonian people at the time the film was shown (see sidebar for subsequent views of two scientists). We are told the group were shown "movies and stills." By the plural word "movies" it appears both film rolls taken by Patterson were screened.

1. The creature was supposed to be a female, yet they all agreed that it walked not merely like a human being, but like a man, a male human being.

2. The creature had a sagittal crest (pointed head) which is not merely an ape characteristic, but a feature of a male ape, and only a mature adult male at that. They reasoned that an uninformed hoaxer would probably make this mistake because gorilla costumes are usually that of male gorillas. Female costumes would not be provided unless specifically requested. *[The inference here being that the hoaxer used a male gorilla costume (i.e., a gorilla is a gorilla) and changed it to a female by adding breasts. Being uninformed, the hoaxer failed to realize that females do not have a sagittal crest. They went on to reason that a large man inside such a suit, padded out in the shoulders and bust, would be shaped like a giant female, but would give himself away (at least to experts) by his male gait, even without the male sagittal crest.]*

3. One physical anthropologist who specializes in bone structure stated that judging from the footprint (note singular term) left by the creature, the toes were too short for the length of the foot. *[I believe the reference here is to a cast of the footprint which would have likely been provided. There are no clear footprints seen in the film showing the creature. However, such are seen in the second film roll and it is possible stills (ordinary photographs) were provided of actual footprints (i.e., the Laverty photographs). Nevertheless, if either were the case I think the plural term "footprints" would have been used.]*

4. A comment was made on the odor Patterson reported; quoting from the source: "What creature could 'smell terrible' to a human nose at a distance of a hundred feet in open air?"

Subsequent Scientific Views

Two of the scientists at the Smithsonian viewing were Dr. Richard Thorington Jr. (who succeeded Dr. John Napier as director of the Primate Biology Program), and Dr. William Montagna, director of the federal primate center at Beaverton, Oregon. They both made their views known some years later. What they stated is as follows.

Views of Dr. Richard Thorington Jr.

…To assessment of the reaction to that historic viewing of the Patterson-Gimlin film:…To them it appeared all too obvious that the pictures were made of a person dressed up in an ape costume, trying to run in an unnatural way. The only person who did not at the time consider this to be the case was Dr. John Napier…let me point out that there are two ways to look at the Bigfoot phenomenon. One is the approach that Dr. Napier espoused that we should keep an open mind and review all evidence to decide whether this is a hoax or a legitimate area of study.

Views of Dr. William Montagna (1976)

Along with some colleagues, I had the dubious distinction of being among the first to view this few-second-long bit of foolishness. As I sat watching the hazy outlines of a big, black, hairy man-ape taking long, deliberate human strides, I blushed for those scientists who spent unconscionable amounts of time analyzing the dynamics, and angulations of the gait and the shape of the animal, only to conclude (cautiously, mind you!) that they could not decide what it was! For weal or woe, I am neither modest about my scientific adroitness nor cautious about my convictions. Simply stated, Patterson and friends perpetrated a hoax. As the gait, erect body, and swing of the arms attest, their Sasquatch was a large man in a poorly made monkey suit. Even a schoolchild would not be taken in. The crowning irony was Patterson's touch of glamor: making his monster into a female with large pendulous breasts. If Patterson had done his homework, he would have known that regardless of how hirsute an animal is, its mammary glands are always covered with such short hairs as to appear naked.

5. Some thought was apparently given to the cost of making a costume for the purpose of a hoax. It appears an estimate of such a cost was provided. The comment (or consensus) was that while an ape man costume would be expensive, it would be nothing like the price quoted. The opinion was that the price quoted was probably for the provision of a mechanical sasquatch.

6. One scientist "allowed the possibility of the film being genuine" (i.e., shows a natural creature), even though he mentioned reservations.

The information provided raises four specific issues concerning the nature of the creature:

Point 1: The unusual walk
Point 2: The sagittal crest
Point 3: The short toes
Point 4. The odor

The first two points were raised at the University of British Columbia screening and have been addressed by John Green in the previous section, They definitely do not detract from the credibility of the creature filmed.

The third point (short toes) has been addressed by Dr. Jeffrey Meldrum. His conclusion on the issue is quoted as follows:

The Bluff Creek tracks (film site tracks) don't really have short toes. They simply appear short at first glance due to a slightly more extensive sole pad at the base of the toes. Closer examination reveals the presence of a flexion crease that marks the position of the hallucialmetatarsal joint at the base of the big toe, which position is consistent with tracks from elsewhere that appear to have longer toes. This also explains the apparent "double ball" feature that is present in a few of the Bluff Creek tracks but not evident elsewhere.

The fourth point (odor) what is believed to be the case here has been addressed by both Dr. Henner Fahrenbach and Dr. Jeffrey Meldrum. Their conclusions are shown in the sidebar.

It should be noted that when the creature in the film was first spotted, according to Bob Gimlin it was only about 50 feet (15.2m away). I believe an odor could be sensed at this distance.

As to the Smithsonian thoughts on the cost of a costume, I do not know what estimate they were given. It is, however, amusing that the prospect of a "mechanical sasquatch" is even mentioned. We don't even have the technology at this time to produce a mechanical sasquatch like the creature seen in the film. The point made is totally absurd. I really can't believe that any Smithsonian scientist would entertain this thought.

Dr. Henner Fahrenbach
Sasquatch Odor

Some sasquatch reports mention of an intense stench. By comparison to the well-documented aroma of excited gorillas ("over-powering, gagging aroma at 80 feet," Dian Fossey), we can speculate that a sasquatch under stress produces this odor from its axillary (arm pit) glands, buttressed by the generally obnoxious body scent of a soiled primate. The frequently mentioned perception of "being observed" and the displayed fear of other animals before a sasquatch encounter might be caused by a yet to be discovered pheromone effect, producing an automatic flight-or-fight response in man and animals.

Dr. Jeffrey Meldrum
Sasquatch Odor

Occasionally, a distasteful pungent odor is experienced in association with a sasquatch encounter. The odor can be rather overpowering and is compared to the smell of rotten eggs, putrid meat, or rank body odor. However, much more frequently, no noticeable odor is detected during an encounter, even at close quarters. A mere 10 percent of the reports accumulated by John Green make any mention of an odor. In his interactions with the mountain gorilla, Dr. Schaller noted an odor described like pungent human sweat, manure, and distant burning rubber. He suspected it emanated primarily from the silverbacks when the group was in a state of excitement. Indeed, the male gorillas have well-developed axillary organs, located in the armpits, comprised of apocrine sweat glands. The same type of glands developed to a lesser degree in humans with the onset of puberty. These can reflexively discharge a strong musky odor in response to fear or threat. Dian Fossey recounted one of her early encounters with a charging silverback gorilla when she approached the group too closely. The onrushing patriarch gorilla stopped just short of her position, but she was hit by a powerful musky odor that emanated from the ape. The function of well-developed ape axillary organs may explain the inconsistent reporting of an associated strong odor during sasquatch encounters.

10. European Scientific Reviews

Although the film failed to get recognition from the scientific community, it continued to arouse interest in North America. In addition to *Argosy* magazine, three other major magazines ran articles on the film story. These magazines were *National Wildlife* (April/May, 1968), *West Magazine* (December 1968), and *Reader's Digest* (January 1969). In March 1969, *Reader's Digest* included the article in their international edition. This action brought the story to the attention of people in many European countries.

In 1971, René Dahinden came to the conclusion that he must look beyond North America for scientific involvement. Scientists in Canada and the United States had been given more that an adequate opportunity to analyze the findings. Certainly, a few intrepid scientific individuals had expressed interest, but no serious plans for detailed study had been offered.

After arranging contacts, Dahinden traveled to Europe in November 1971. He visited England, Finland, Sweden, Switzerland, and Russia. Although scientists in these countries were somewhat more open-minded than those in North America, their conclusions were basically the same, notwithstanding the "qualified" findings of Dr. Donald W. Grieve, a biomechanical expert in London. He concluded that if the film speed was set at 16 or 18 frames per second, a human being could not duplicate the creature's walking pattern. If the speed was 24 frames per second, the pattern could be duplicated. As Patterson did not know the filming speed, we were now faced with yet another tantalizing question.

A real glimmer of hope, however, emerged at the Russian Central Scientific Research Institute of Prosthetics and Artificial Limb Construction. Here, scientists concentrated on the creature's movement or locomotion. They concluded that the creature was extremely heavy. Great weight was indicated by how the creature's arms swung and how its knees bent when its body weight came onto its feet. They stated that bulk can be simulated, but not massive weight. They were also impressed with the creature's apparent muscle mass. The institute agreed to do some biomechanical work on the film, but never proceeded. Furthermore, it never provided any official written opinions on its preliminary conclusions, other than a letter to the USSR Committee on Cinematography stating that, "The film contains sufficiently clear frames of the walk of a manlike creature, a detailed study of which would undoubtedly be of serious scientific interest."

Nevertheless, hominologists Dmitri Bayanov[1] and

René in Red Square, Moscow. (Photo, D. Bayanov)

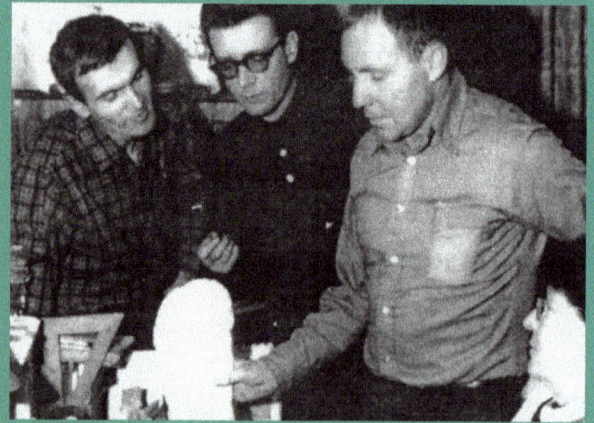

René showing a film-site cast to hominologist friends in Moscow. To his extreme right is Igor Bourtsev, then Dmitri Bayanov. In the foreground is Dr. Marie-Jeanne Koffmann. (Photo, D. Bayanov)

Igor Bourtsev[2] expressed great interest in the film. These men were members of the Smolin research group, which is affiliated with the Darwin Museum in Moscow. The group was formed in 1960 to research relic hominoids in the Soviet Union. Dahinden provided a copy of the film and footprint casts to Bayanov and Bourtsev, and they commenced the most intensive film analysis up to that time.

In their study, Bourtsev addressed the filming speed (frames per second).[3] The camera had the capacity to take films between 16 and 64 frames per second. Actual setting positions on the speed dial were at 16, 24, 32, 48 and 64 frames per second. Bourtsev analyzed the film's vertical oscillations which revealed that the film was taken at 16 frames per second. According to Dr. Grieve's conclusions, therefore, the creature was not a hoax.

Bayanov also made an important discovery. On a film frame showing the sole of the creature's right foot, Bayanov found a detail that corresponded with the plaster cast. This discovery firmly connected the creature filmed with the footprints.[4]

Furthermore, both Bourtsev and Bayanov worked on arriving at an estimated weight for the creature. The team concluded that the creature weighed *no less than 500 pounds.*[5]

These findings, and other findings supporting the authenticity of the creature, were published in Russian (*Science and Religion* magazine, 1976, #6) in an article by Bayanov and Bourtsev entitled, "The Mystical Biped." In North America, their findings were published in the book, *The Sasquatch and Other Unknown Hominoids*, edited by Vladimir Markotic (1984).

During this time, Alexandra Bourtsev, Igor's wife at the time, sought out Dr. Dmitri Donskoy, Chief of the Chair of Biomechanics at the USSR Institute of Physical Culture, and asked him to examine the film. Donskoy agreed and provided a detailed report in which he concluded, "such a walk as demonstrated by the creature in the film is absolutely non-typical of man."

DETAILS AND ANALYSIS

1. Dmitri Bayanov

Bayanov was born in 1932 at Moscow. He graduated from a teachers' college in 1955. He originally worked as a regular teacher and later made a living as a Russian-English translator. Highly interested in hominology, Bayanov became an active member of Relic Hominoid Research Seminar, Darwin Museum, Moscow. He was chairman of this group in 1975 and subsequent years. He is an original member of the Board of Directors of the International Society of Cryptozoology. Bayanov translated into Russian the *Argosy* magazine article by Sanderson (1968). The article was published in the Russian popular science magazine, *Znanie-Sila*.

Igor Bourtsev (left) and Dmitri Bayanov in 2002. (Photo, Daniel Perez)

Dr. Dmitri Donskoy. (Photo, I. Bourtsev)

Dmitri Bayanov and Dr. Jane Goodall, 1972. Jane has expressed that she believes in the existence of sasquatch or bigfoot. She has spoken directly to native North Americans and is truly impressed with their sincerity as to these creatures. (Photo, I. Bourtsev)

2. Igor Bourtsev

Bourtsev was born in 1940 at Samarkand, Uzbekistan, USSR. He graduated as an engineer in 1963 from the Aviation Institute at Moscow. In 1977, he completed a postgraduate course in history and sociology, obtaining the degree of Candidate of History. He became involved in hominoid research in 1965. He has taken part in many expeditions into the Caucasus and Trans-caucasus, gathering information on alleged unusual hominids in those regions. On three separate occasions he endeavored, without success, to find the grave of Zana, an alleged hominid female who died in the late 1800s.

3. Film Speed Research—Bourtsev

Patterson took some of his movie while running. The jerking and shaking of his movements are reflected in the film. Bourtsev took the vacillation of images on the film and related them with Patterson's steps and movements. In this way, Bourtsev claimed it was possible to determine the filming speed. He stated that it can be seen that sometimes one oscillation of the camera corresponds to 4 frames, or 4 oscillations to 16 frames, or 6 oscillations to 24 frames. Every oscillation corresponds to one of Patterson's steps. If the filming speed was 24 frames per second, this means Patterson was taking 6 steps per second, which is impossible. The best athletes running 100 meters in competition take only 5 steps per second. Patterson's maximum pace was 4 steps per second, which equates to a filming speed of 16 frames per second.

4. Sole of the Creature's Right Foot Detail

Bayanov discovered that the film showed a double ball on the sole of the creature's right foot. This foot is very clear in Frame 61 of the film. The frame shows the creature from the back, in walking motion, with its right foot lifted. Bayanov found the same double ball configuration in a plaster cast taken of the right foot footprint at the film site.

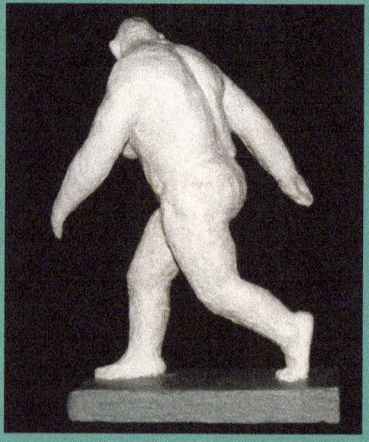

5. The Weight of the Creature

Bayanov and Bourtsev established a minimum weight of 500 pounds by comparing the creature's visual bulk to a human male. They established that the creature was 2.25 times larger than the human. They took the known weight of the human and multiplied it by 2.25 to arrive at the final figure. This calculation, of course, assumes that the creature is built the same as a human being. That is, it has the same bone and flesh density. It is reasonable to assume that the creature's density would not be *less* than a human being. A *minimum* weight of 500 pounds is therefore substantiated.

This sculpture of the creature as seen in Frame 352 of the film was created by Igor Bourtsev. It was sent as a gift to René Dahinden after his visit to Russia in 1971.

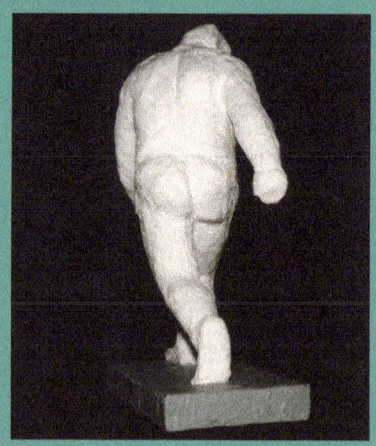

More sculpture images are at the following link:

http://forum.hancockhouse.com/images/articles/20061102142036149_1_original.jpg

11. Current Findings and Insights

The NASI Report

From 1995 to 1998 an intensive scientific analysis of the Patterson/Gimlin film was undertaken by Jeff Glickman, a certified forensic examiner. The analysis was originally commissioned by Peter Byrne, director of The Bigfoot Research Project. During the course of the analysis (May1997), The Bigfoot Research Project was replaced by the North American Science Institute (NASI) with Glickman as director. In June 1998, Mr. Glickman released a research report entitled "Toward a Resolution of the Bigfoot Phenomenon" (commonly referred to as the NASI Report). The report findings are highly positive that the creature filmed was a natural animal. Mr. Glickman declares the following in his closing statement:

"Despite three years of rigorous examination by the author, the Patterson-Gimlin film cannot be demonstrated to be a forgery at this time."

Although work on the analysis commenced in 1995, neither René Dahinden nor I were aware that the analysis was already in process when approached by Peter Byrne in June 1996 with his analysis proposal. At that time, Byrne told us that his plan was to produce a scientific report and provide it free of charge to all major universities and major research organizations worldwide. We agreed with Byrne's plan and an agreement was effected in September 1996 to allow The Bigfoot Research Project to publish photographs of film frames.

Certainly, it was possible that the analysis could indicate, or even prove the creature was a hoax. The Bigfoot Research Project, however, had fully considered this possible eventuality. The following is from Peter Byrne's notes, that he provided to me in 1998.

If our study should expose the footage as a fabrication, with its subject simply a large person in a man-made costume, then, we felt, so be it, let it join the Piltdown man and others of his ilk. If on the other hand, the study produced evidence that supported the reality of the footage, then among other things, we would be opening the door for scientific inquiry to be conducted by others with expertise in related fields; others who could use our findings as stepping-stones towards further research and discovery. We could, in what we proposed to do, either put an end to the twenty-nine years

René, Peter and me at René's place after discussing the contract for the use of film frames, June 7, 1996. (Photo, Peter Byrne)

Front and back covers of The Bigfoot Research Project brochure. It was a nice publication, and for me these were exciting times—it seemed only a matter of time…

The entire brochure is shown at the following link:

http://forum.hancockhouse.com/media/brochuretbrp.pdf

life span of one of the most sophisticated hoaxes of the twentieth century or, conversely, open ourselves, and the world of science to a veritable ocean of new knowledge in the fields of primatology and anthropology; knowledge that could, among other things, shed new light upon the dark areas of our human origins.

Unfortunately (in my opinion) in late 1997 NASI changed direction on the master plan. It solicited for paid memberships in NASI and worked towards getting the report, when finished, into a major scientific magazine. Both objectives failed. While the report was eventually featured on a web site, it was never officially published. Within a few months after the report was released (June 1998), Mr. Glickman left NASI and a short time later NASI folded (late 1998 or early 1999). It is my understanding that the backer of the organization withdrew his support. It is this occurrence that I find truly amazing. This person had invested many thousands of dollars into the project and the final results **as we know them** were as **favorable** as can be expected for this type of study.

Artwork from The Bigfoot Research Project brochure depicting the Patterson/Gimlin experience.

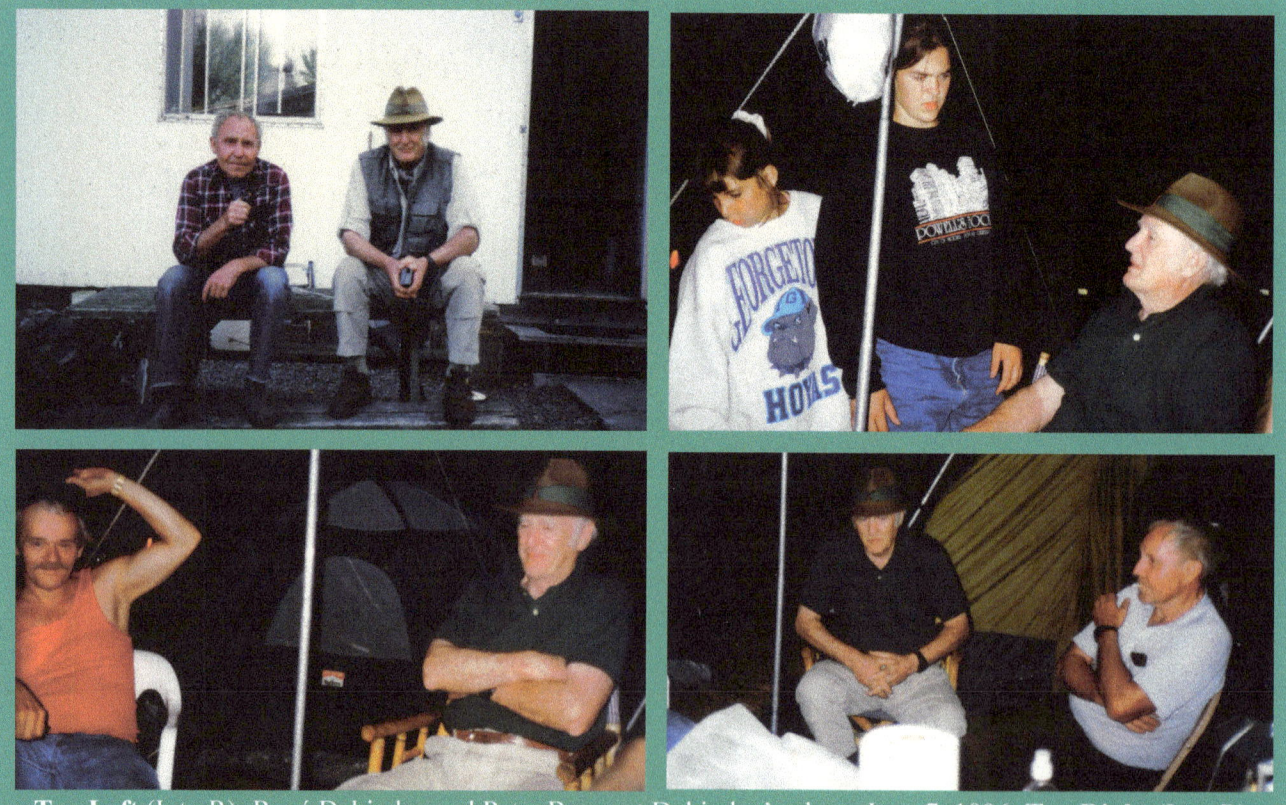

Top Left (L to R): René Dahinden and Peter Byrne at Dahinden's place, June 7, 1996. **Top Right** (L to R): Mallory Webb (Ray Crowe's step-granddaughter), Rara Byrne (Peter's daughter), Peter Byrne. **Lower Left** (L to R): Larry Lund and Peter Byrne. **Lower Right** (L to R): Peter Byrne and René Dahinden. The last three photos were taken at Ray Crowe's Carson, Washington camp-out, summer 1996. This camp-out, to my knowledge, was the last occasion during which Byrne and Dahinden were photographed together.

As far as I know, there was no consideration given to the work by any major scientific organizations. I did not even see a newspaper article or television news announcement on the results of the project. Despite a controversial creature weight estimate, I still believe we would be much further ahead if the original plan formulated by Peter Byrne for the work had been followed.

Sasquatch: Legend Meets Science

In 2003, some professional analysis of the Patterson/Gimlin film was documented in a video entitled, *Sasquatch: Legend Meets Science,* produced by White Wolf Entertainment. The way in which the creature walks was thoroughly analyzed, with the conclusion that its walk was totally different from that of a human being, and could not be duplicated by a human being. This same conclusion was reached by Jeff Glickman and was stated in the NASI report. Moreover, there is a large lump on the creature's right thigh, which may be associated with its musculature or could be a hernia. It is not reasonable that a detail of this nature would be seen in any sort of fabrication. (Note: A book by Dr. Jeff Meldrum also titled *Sasquatch: Legend Meets Science* was released in 2006 — see sidebar).

The Sarmiento Report

In March 2002 Dr. Esteban Sarmiento, then with the Museum of Natural History, New York, completed his analysis of the Patterson/Gimlin film. His basic conclusion was that the creature seen was within human proportions and was probably human. Although he falls short of stating that the creature was a hoax, he implies this possibility. The closing statement in his report provides the following information.

> Only through bio-molecular studies and/or dissection of living or cadaver specimens, and proof of parallelisms, could this individual be called anything but *Homo*.
> …Bigfoot's movement, especially its neck and trunk movement, indicate it is more or as closely related to modern humans as some of the fossil taxa within the genus *Homo*.

Non-Professional Findings

Non-professionals continue to probe the Patterson/Gimlin films for clues supporting the creature's reality or otherwise. Marlon K. Davis of Mississippi, using fairly sophisticated equipment, has recently pointed out additional movements (in addition to those pointed out by NASI) in the creature's flesh and muscles, plus foot contours, and other surface details, that don't support artificial coverings. Notwithstanding testimony and hearsay, there is far more evidence supporting the filmed creature's reality than otherwise.

The video *Sasquatch: Legend Meets Science* and the book of the same name by Dr. Jeffrey Meldrum provide the most current testimony to the reality of the creature filmed by Patterson and Gimlin. Dr. Meldrum's summary is as follows:

"Those who have given the film the most thorough consideration have, with few exceptions, concluded that it is probably authentic or that its authenticity cannot be readily eliminated given the limitations or image quality or want of a definite scale."

MORE ON DAVIS' FINDINGS: Davis believes the creature filmed was a human of some sort. The following link provides some insights:

http://forum.hancockhouse.com/article.php/20061214194131628

12. The Major Issues

The Patterson/Gimlin film is *one* piece of evidence supporting the existence of bigfoot. It is not the only piece of evidence. However, the film is certainly the *most important* piece of evidence to date. If the film were proven to be a hoax tomorrow, this would not herald the end of belief in bigfoot. Nevertheless, this finding would be a setback in the field of bigfoot research. Taking all aspects of the film into consideration, I consider the following the **major issues:**

A. Absence of Additional Photographic Evidence

With thousands of sighting reports, one would think someone would get a good picture of the creature. It is possible someone has, although I doubt it. I believe that if there were any good photographs or films/videos "out there" they would surface very quickly.

Why have we not been able to get more reasonably clear pictures of the creature in the last forty years? Cameras have come a long way since 1967. There are now many more people carrying them and they are exceedingly more efficient.

We can, however, certainly rationalize lack of film evidence. In the first place, one must have a camera when a sighting occurs. Then there is the problem of positioning and focusing a camera on split-second notice. Moreover, further complications arise because people are so overwhelmed when they see the creature.

Roger Patterson was an exception because he went out specifically to get bigfoot evidence. He would have thought about an encounter many times, and therefore conditioned himself in this regard. I will also mention here that Patterson was on horseback. Horses allow for minimal intrusion in wilderness areas (far superior to motorized vehicles or even walking). Perhaps we can learn something here.

Nevertheless, I am quite sure if another bigfoot investigator saw a sasquatch, he or she would be able to get a good photograph, given there was enough time. This thought brings us to another point: there are very few bigfoot investigators out in the bush looking for the creature. Effectively, the sasquatch has not been found (and photographed more/better) because few people are actually looking for it. The number of hardy individuals we have "beating the bushes" is so insignificant it hardly counts. Indeed, most researchers are sent out (or go out) "after the fact." In other words, they are responding to a sighting by a non-interested person (tourist, farmer, hiker, camper, and so forth). In effect, we are sitting back waiting for one of the researchers to "luck out."

There is certainly no shortage of *perceived* bigfoot images. People see something at a distance that looks like a bigfoot head and part of its body and take a photo. Also (most often) they see something that looks like a bigfoot in the background after they get their photographs developed. Shown above is what is happening here. The first photo is an image registration of the three head segments seen in the second photo. I simply stood in front of the books and lined up the head parts with my camera. Leaves, branches, shadows sort of flow together and produce images. When one looks again, the image is gone because of movement of the components, viewing position or lighting changes. Bigfoot is everywhere if one has a good imagination.

B. Lack of Major Investigator Firsthand Findings

The only major bigfoot researcher in the West, other than Roger Patterson and Bob Gimlin, who actually saw a bigfoot was Bob Titmus. None of the other major "Western" bigfoot investigators (René Dahinden, John Green, Peter Byrne, Grover Krantz, Jeff Meldrum, Richard Noll, or Thomas Steenburg) in all their many years of experience saw one of the creatures. Furthermore, in most cases, footprints have not been found by these investigators "firsthand." In other words, they found the prints themselves, not as a result of someone else reporting prints and then going to investigate the report. Certainly the major researchers independently found *additional* prints. That is, they responded to a report and upon investigation found prints that had not been observed by the person who made the report.

Once in one of my long sessions with René Dahinden, we talked about these issues. After we had exhausted every possible angle, one of those "moments of silence" occurred. As I gathered my belongings together, Dahinden said, very quietly and deliberately, "You know, I have been searching the bush for over forty years, I think that says something." This was the only statement Dahinden made that I can recall whereby he expressed some marginal doubt on the existence of the creature. Nevertheless, that one man alone would **expect** to find the creature (as he did) is pushing the limits of probability—much like a lottery win. By the same token, he also told me that he seldom saw bears or other large mammals during his solo expeditions. Remarkably, in all that time, he saw only one cougar.

C. Missing Evidence

There is an irony that seems to attach itself to all "mysteries." For some inexplicable reason, critical evidence or information is usually missing. Generally such material is lost or destroyed as a result of some unusual event. The Patterson/Gimlin film is no exception. At this writing, we have still not been able to access the original film roll; the second film roll is effectively lost; and we do not know who specifically developed the films.

D. Lack of Absolute Indicator for Creature Measurements

There is no *absolute* or *direct* indicator for the estimated height of the creature in the Patterson/Gimlin film. Measurements are all based on some "questionable" factor (casts, camera distance, photo registration, etc.). Knowing the height of the creature is highly important in determining if it could have reasonably been a human being in a costume. The creature's height estab-

Given that the creature filmed was about 87.5 inches tall, and Roger Patterson was about 66 inches tall, including his boots and cowboy hat, then what is shown here is an approximate comparison of the two. The silhouette of Roger is from the photograph of him holding casts at the film site.

If the creature was just 72 inches tall, then this approximate comparison would apply. Either way, we can see that the creature's size would have impressed both Patterson and Gimlin.

NOTE: I have simply worked out relative percentages. This is not a "scientific" presentation.

lished by NASI *virtually* places it outside the human range. Other findings, however, place it totally within the human range, although at the high end (well over 6 feet tall).

E. Inability to Detect Possible Hoax Indicators (Fasteners etc.)

According to Jeff Glickman and Dr. Henner Fahrenbach, possible hoax indicators on the creature in the Patterson/Gimlin film, such as zipper pulls, rivets, buttons, etc., would not be visible. The image portion of a film frame in the Patterson/Gimlin 16 millimeter film is 7.6mm high and 10.29mm wide. The bigfoot creature seen in the actual frames is about 1.2mm high. While the frames can be greatly enlarged (projection or photographically), it is mathematically impossible to see any credible detail less than about 3.6mm in diameter on an image of the creature that is a maximum of 96mm high. **What can be seen with the naked eye on this image is all the credible detail available**. If one takes a magnifying glass and sees additional "information," such has absolutely no credibility. The adjacent photograph illustrates the process.

The only exception to this rule **may** be observance of exactly the same detail on consecutive frames, or a long series of related very small images (such as seam, threads, or zipper teeth). Here, however, as the creature is moving, it is unlikely the images would be successively identifiable or would register (i.e., be seen in a series).

Given the limitation, even if the creature were a fabrication it would be virtually impossible to detect meaningful hoax indicators, other than the improbable exceptions stated. This problem is further compounded by the fact that hoax indicators would probably not be visible in the first place. Costume makers (both professional and amateur) are careful to conceal these details. However, *for what it is worth*, sophisticated computer analysis by NASI of the best film frames did not uncover any hoax indicators. I say, "for what it is worth" because it is impossible to uncover something that cannot be seen in the first place. The only thing NASI **might** have been able to find was a "series image" as I have related—a very, very tough call.

G. Height and Weight Discrepancy

Our "professional" estimates of the creature's walking height range from 5 feet 8 inches to 7 feet, 3.5 inches. Weight estimates range from 190 pounds to 1,957 pounds. The spread in both ranges is far too great to rationalize. Generally, everyone has "agreed to disagree" on the subject of height and weight.

The creature seen here is 96mm high from ground level. The white dot on its shoulder is 3.6mm in diameter. Anything smaller than the dot **seen with the naked eye** does not have credibility.

A detailed paper on film resolution is provided at the following link:
http://forum.hancockhouse.com/article.php/20060324230946666

For the Weight Watchers

When Patty's weight of 1,957 pounds was determined by the North American Science Institute (NASI), most researchers could not believe the figure, so I decided to do my own calculation using the weight of water as a basis. Given she was 87.5 inches tall, I carefully calculated the sizes of her various parts, and came up with 2070.78 pounds—very close to the NASI figure. There was enough water to fill 4.6 forty-five imperial gallon drums. I suppose I must have made a mistake somewhere along the line, however, to this day I wonder, but perhaps understand why NASI never relented on the weight it determined.

13. The Major Hoax Claims

At the time of the publication of this work, there have been six (6) major claims that the Patterson/Gimlin film was, or probably was, a hoax. I discuss each in turn. It might be noted that I do not consider the Ray Wallace claim as a major claim. It is too ridiculous to consider in this category.

1. The John Chambers Connection

The John Chambers connection was a "natural." Chambers had designed the ape heads/faces for the movie *Planet of the Apes,* which was released in 1968, the year following Patterson and Gimlin's experience at Bluff Creek. The apes in the movie were very convincing, but totally different from the bigfoot seen in the Patterson/Gimlin film. It was, however, reasoned by some people close to Chambers that he could have, or might have, created the Bluff Creek bigfoot. Apparent rumors made the rounds in Hollywood circles and were picked-up by writers. Why Chambers would have created the creature, and who would have compensated him was never speculated. The controversy flared up and died down over the years, but continued to "get press."

Chambers was finally requested to be interviewed on the issue in 1997. At this time, he was in a retirement home. A spokesperson for Chambers replied to the request informing that Chambers was not available for interviews, but went on to state, "He said that he did not design the costume." This statement was taken to possibly mean that someone else designed "the costume" and Chambers made it. Another veteran make-up artist and friend of Chambers added fuel to the fire by stating that Chambers would never admit it if he were involved.

Fortunately, in that same year the whole issue was laid to rest by Bobbie Short, a registered nurse and bigfoot researcher. Bobbie was granted a personal interview with Chambers and received direct answers to her questions. Not only did Chambers deny any involvement with the Patterson/Gimlin film, but in his opinion, **neither he nor anyone else could have fabricated the creature seen in the film.** Chambers stated that he was good, *but not that good.* He admitted that he was aware of rumors concerning his involvement in the film. He never took steps to set the record straight because it was "good for business." One final note—Chambers had not even heard of Patterson or Gimlin prior to October 1967. He had never heard of Al DeAtley.

2. The Harry Kemball Fiasco

The Harry Kemball fiasco was far less credible than the Chambers connection. In a letter to The *'X' Chronicles*

Bobbie Short in 2004. Bobbie continues to be an ardent and skillful bigfoot researcher. Her website *Bigfoot Encounters* is one of the main sources of information on the creature, ranging from historical records to current findings. Bobbie herself had a sasquatch sighting in 1985.

John Chambers in a photograph taken by Bobbie Short on October 26, 1997 during her interview with him. He died on August 25, 2001 at age 78. He was a remarkable man who will be long remembered.

Roger Did Have Artistic Talent, But...

This sculpture by Roger Patterson demonstrates that he could both draw and sculpture. Could he have made a reasonably good bigfoot head mask? I think so. Could he have created the entire body of the creature with moving muscles and flexible feet and hands? I don't think so. If he had that level of talent, then I'm sure he would gone to Hollywood long before he took up the search for bigfoot.

Bobbie Short's web site:
http://www.bigfootencounters.com

dated May 14, 1996, Kemball, a director and screenwriter at Golden Eagle Productions, stated he saw Patterson and friends put together a film hoax. Kemball claimed he witnessed this event at the CanaWest film facility in North Vancouver, British Columbia. In his <u>own</u> words (*exactly* as he wrote), Kemball stated:

> I was in the CanaWest 16mm Film Editing Room in 1967 when Roger Patterson and friends put together his BigFoot Hoax on 16mm film. They all laughed and joked about the rental of the Gorilla Costume and the construction of the Big Feets. One of his extra tall buddies played the Role of BigFoot. They carefully chose muddy ground so that the foot prints would expand. They carefully shot it on 16mm Kodak EF High Speed Color Positive film stock and when the film is force processed in "Hot Soup"– the film grain is enlarged to add to the sense of mystery. They added a shaky camera zoom with the right amount of "out of focus" to complete the deception. It amazes me that the frame you published on the front page of your Jan/Feb issue is the same frame that all the Media World-Wide has used over the last 29 years. I sincerely hope that Patterson isn't getting paid for this nonsense! As a graduate in comparative anatomy studies this creature does not and has never existed.

The *'X' Chronicles* admittedly misread the letter and stated in their release that it was one of *Kemball's extra tall buddies* who played the role of bigfoot. A newspaper tabloid got hold of the story and naturally added more misinformation. The tabloid stated that a person (Kemball) had come forward *who claims to have helped create the film.* Each of these details, of course, gave unwarranted credibility to Kemball's claims.

Pete Byrne of The Bigfoot Research Project contacted Kemball, which resulted in more ridiculous claims by the latter. Kemball incorrectly identified the type of camera Patterson used. He also stated that Patterson had made a deathbed confession to the effect that the film was a hoax. Yet it appears that at the time Kemball wrote the letter, he was unaware that Patterson had died 24 years earlier. When asked why he did not come forward with his claim some 29 years ago, Kemball's reply was surprising. He stated that in the little town of Cranbrook, British Columbia, where he lives, there had been a recent upsurge in crime. There was even a police standoff that apparently shocked the residents of the sleepy town. Kemball reasoned that hoaxes cause crime

For the Record

John Chambers did make a bigfoot statue that was used in a parade float in the mid 1960s. As with his *Plant of the Apes* consumes, the statue has little resemblance to the creature seen in the Patterson/Gimlin film. I find it amusing, however, that the statue was 7 feet, 4 inches tall—very close to the 7 feet, 3.5 inches determined for the Bluff Creek creature.

There have certainly been many films about bigfoot, and with new technology, some of the costumes are very convincing, but none can compare to the Patterson/Gimlin film.

Kemball Quirks

The Kemball disclosure was a bit of a surprise because this man is quite well-known in movie production circles. There can be no doubt that he often goes to film processing facilities and has a lot of knowledge in the movie-making business. He certainly does not appear to be the kind of person who would come up with a story for publicity purposes.

That Patterson was in Canada prior to the filming at Bluff Creek appears evident by his own admission. However, we don't know the time frame.

I do think Kemball saw and heard something, as I point out, and it would be consoling to know exactly what happened.

One thing that highly surprises me is that Greg Long does not mention the Kemball incident in his book, *The Making of Bigfoot.* It was all over the Internet (I posted Peter Byrne's findings) and Kemball was on at least one radio program talking about what he experienced.

I do not take this lightly, as Long has demonstrated his research abilities and I don't think he overlooked Kemball. I will guess that Long sorted out the matter and found that it did not have substance, so simply made no reference to it. Most certainly, he would have played that card if there were a ghost of a chance it would take a trick.

so he released the information as a crime-fighting gesture.

Nevertheless, it is reasonable to assume that Kemball based his claim on something. As it happened, René Dahinden (who was five feet six inches tall) and John Green (who is six feet, 3 inches tall), visited the CanaWest laboratory in January 1968 to have some work done on the Patterson/Gimlin film (enlargements, etc.). It is very possible staff at the laboratory joked about the film and were overheard by Kemball. Also, it is very likely Kemball mistook Dahinden and Green for Patterson and his "extra tall buddy."

There is, however, one minor reference that indicates Patterson was in Canada prior to October 20, 1967. This reference is in *The Times-Standard* article published on October 21, 1967. The article states: "This year he *(Patterson)* has been taking films of tracks and other evidence all over the Northwestern United States **and Canada** for a documentary."

The 'X' Chronicles suggested solution to the whole issue was to have Bob Gimlin put into a hypnotic state on a television show whereupon he would be asked questions relative to the film. He would then be submitted to a polygraph test and a psychological stress evaluation test. Needless to say, this process did not take place, and the Kemball fiasco died a natural death.

3. The Clyde Reinke Disclosure

On December 28, 1998 a television documentary entitled *The World's Greatest Hoaxes* was aired. The Patterson/Gimlin film was discussed along with other "unexplained" subjects. Clyde Reinke of American National Enterprises claimed that Roger Patterson was employed by this company to participate in the filming of a fabricated bigfoot sighting. We are led to believe that the resulting film was the famous Patterson/Gimlin footage. Reinke stated that as personnel director for American National, he signed Patterson's paychecks. Reinke identified a man, Jerry Romney, who was alleged to have acted as the "creature." Romney was interviewed but he flatly denied any involvement.

Nevertheless, all of this information is totally misleading. While Roger Patterson was certainly associated with American National Enterprises, this association did not take place until 1970—about three years *after* he obtained his footage of the creature. American National wanted to get their own movie of the creature for a specific production and hired Patterson for this purpose. In 1971, American National abandoned the project and decided to use Patterson's footage for their production.

There are possibly two reasons why American

Second Roll Images

There is no doubt that the producers of *The World Greatest Hoaxes* had access to all or part of the second Patterson/Gimlin film roll. Right at the beginning of the documentary, we see footage of Patterson making a cast at Bluff Creek. Printed information is superimposed on the images. I know the footage was from the second roll because the images match those I have that were from that source.

Film Follies

Patterson's subsequent agreement to allow American National Enterprises (ANE) to use his film apparently included provision of the original (actual) film he had taken at Bluff Creek. I can only guess that they wanted the original roll because this would provide the best resolution for reproduction. It would appear to me that the original film should have been returned to Patterson right after it was duplicated, however, it was not. When he died two years later, there was apparently no follow-up done to get the film back.

Subsequent litigation involving the film ANE made, occasioned by René Dahinden, created a legal quagmire with the results that the film will not be provided for inspection without more litigation. The Florida lawyer who we believed had the film would not even simply put it on the table for Peter Byrne who went to the company and asked to inspect the film in the presence of their people. The lawyer told Byrne he did not have the film. However, this appears to be a lie as Mrs. Patterson has told us that a Florida lawyer does have the film. Unless Byrne went to the wrong lawyer, I have no explanation.

I know that René Dahinden, who owned 51% of the film, tried very hard to get access to it. He provided me with correspondence in this regard sent to Mrs. Patterson's lawyer. However, he was not able to do so.

For certain, it is important that the film be properly stored (appropriate climate) or it will deteriorate. If it is now simply stashed away in a safe or cabinet, then it is likely it has already significantly deteriorated.

National did not initially choose to use Patterson/Gimlin's movie. Firstly, they thought they could get their own footage (which appears to indicate they thought the film was genuine). Secondly, they may not have been able to reach an agreement with Patterson for the film rights at that time. (The rebuttal information presented here was provided by John Green **who was also retained by American National Enterprises in 1970 to assist the company with its project**.)

4. The Murphy/Crook "Bell" Issue

This claim involved research by your author that revealed what appeared to be an unusual bell-shaped detail in the creature's mid-section. As the same detail in my opinion could be reasonably identified on several film frames, it was deemed to have credibility, and could be a possible hoax indicator. A number of sasquatch researchers were informed of the finding in September 1998 (see sidebar).

One researcher, Cliff Crook, of Bothell, Washington, gave the detail full credibility as a hoax indicator, and I concurred that this was a possibility. We both searched to identify the detail but nothing was found. Crook then asked to report the find to the media, and I informed him that such would be "his call." I talked to only one (1) reporter, and simply answered a few brief questions. Whatever the case, the *Associated Press* got the story and it went worldwide.

After the first press release on November 29, 1998, analysis of the find was then performed by Dr. Henner Fahrenbach who concluded that the detail was simply "photographic noise." Nevertheless, both Cliff and I continued extensive research to identify the detail. Absolutely nothing was found that even resembles the detail. While I still maintain that the detail exists, I believe it is just debris of some sort caught in the creature's hair.

In my opinion, the prominence given this issue by the media was hardly justified. I had envisioned only an article in a local Bothell, Washington, newspaper. However, it appears the press was looking for something different to report, and bigfoot is always an attention grabber. I am told the report (different versions) ended up in some 148 newspapers throughout the world. It is unfortunate this type of coverage is not given findings that support bigfoot existence. Regrettably, bad news is always given priority with the media.

Note: That is the story, and I respectfully request that book reviewers, skeptics, and others not use the newspaper reports (some are ridiculous) to draw conclusions.

The Report on the "Bell"

Upon determining that there might be an anomaly in the film, I issued a report with photographs, providing a detailed account of the finding. I closed with the following statement:

"I sincerely hope I have wasted my time on this analysis and have done nothing but assign dubious credibility to photographic "noise." In releasing this report, I do not wish to give more ammunition to bigfoot skeptics. By the same token, however, if there is something wrong with the film, I do not want this information concealed."

I sent the report to five major researchers and then followed up with two additional reports on other subjects concerning details observed. Two other researchers asked for the first report and it was provided.

The BIG Difference

Words are cheap. They are simply an expression of thought or perception. The only way one can prove they describe something that really happened is to find some physical or photographic evidence to support them. The reason the Murphy/Crook "bell" issue received so much attention is because the claim was based on something that could be physically seen in the film frames. That it was not what it appeared to be is immaterial—this is a matter of interpretation.

The only way the Patterson/Gimlin film, can be debunked is by the film itself. In other words, the film must provide the evidence. As the "bell" seemingly met this requirement, the finding automatically got top attention

All other major hoax scenarios are based on hearsay or testimony, and as such are immediately suspect.

I will mention here that even if someone were to produce a costume, or part of same, that appeared to match the creature in the Patterson/Gimlin film, it would be impossible to prove that it was the same.

5. The Long Shot

In March 2004, a book by Greg Long entitled *The Making of Bigfoot, The Inside Story* (Prometheus Books) was released. Long had been working on this book for some time, and it was common knowledge in the bigfoot community that he was out to debunk the Patterson/Gimlin film. I was alerted to his activities by Bobbie Short. I met Long at the Celebration of Life ceremony for René Dahinden in April 2001. He asked if I would provide him with material related to the film, knowing that I had an unpublished work entitled *Circumstantial Evidence: The Patterson/Gimlin Film* (essentially this work, although not as complete and accurate). I declined to provide information, stating that I was working on my own book.

When I had effectively completed *Circumstantial Evidence*, I decided not to publish it because it contained too much hearsay. It is both unfair and unethical to form opinions of people and their activities based on perceptions. Indeed, in recent years, I have been referred to in books that state information based on hearsay, and what is said is totally incorrect.

Greg Long was far less discriminating. He contacted many people in the Yakima, Washington area who knew Roger Patterson, getting both positive and negative information on his character and activities. Using the negative information, Long wrote a massive character assassination on Patterson (even belittling the man with cutting remarks and innuendo). In this way, he sets the stage to substantiate a "man in a suit" claim.

Some of the history related to the book appears to be that in November 1998 through January 1999, the Murphy/Crook "bell" issue rattled through Yakima County. People who knew Patterson, and considered the film a hoax, came forward with their opinions, which were published by the *Yakima Herald Republic* on January 30, 1999 (article by David Wasson). None of the people had any hard evidence; they just had recollections of Patterson and events in 1967. In the same article, we learn that Zilla attorney Barry M. Woodard, "confirmed" on January 29 that he was representing a Yakima man who claimed to be the person who wore the suit in the Patterson/Gimlin film.

On January 30, 1999 (early morning), I received a telephone call from the Associated Press. As soon as the person identified herself, I said that I did not wish to talk to the media. She said she was not calling for an interview, and then stated, "I have something you will want to hear." She thereupon informed me that a man claiming to be the person in the "costume" had contacted Zilla attorney Barry M. Woodard the previous day

A paper containing the information I provided in *The Bigfoot Film Controversy* is at the following link:

http://forum.hancockhouse.com/media/hoaxmaterial.pdf

The Heironimus Comparison

A detailed comparison of Bob Heironimus and the creature seen in the Patterson/Gimlin film showed that Heironimus' arms were too short and his legs were too long. It was also seen that Heironimus' eye level would have been situated very low in any sort of costume headpiece. An actual film frame was used for the analysis which I provided in the book *The Bigfoot Film Controversy*. (see Link below). None of this, however, completely rules out a costume of some sort (or Hollywood-style makeup) worn by a tall/large man. It just rules out that Heironimus was that man. There are a number of other points that further support the exclusion of Heironimus, and still others that exclude a costume of any sort. In general, the probability of a human outfitted in some way to make him the creature in the film is extremely low.

In the following comparison, I used one of Brenden Bannon's images *(not an actual film frame, and a different image from that previously used)* to "ballpark" a comparison with Heironimus (drawing is from a photograph). I used the right leg from the knee down of each image as a basis for the relative sizes. The creature's leg is more angled than Heironimus' leg, but I think if it were at the same angle, the knees would reasonably match. This is hardly "scientific," but might provide a bit of an insight. We can see that considerable "bulk" would be needed for a costume of some sort, and that it would have to have been extremely well tailored.

On the negative side, I find it very coincidental and surprising that Heironimus was from Yakima and was a friend of both Patterson and Gimlin. All I can say here is that this was a good move on someone's part. Naturally a friend would be given more credibility than a total stranger.

and that this person had passed a lie-detector test. She more or less credited me with "flushing out" this individual.

I believe Long started his intense research on the film during November 1998. The "bell" issue appeared in the press November 29, 1998, so was probably a part of his inspiration (or was his inspiration) to move forward with his investigation. Long tells us he knew the identity of Woodard's "client" in December 1998. If so, then this was before the "client" contacted Woodard. Over five years later, (March 2004) we are told that this person is Robert Heironimus, a Yakima resident, who knew both Roger Patterson and Bob Gimlin.

It appears evident Heironimus thought a "book deal" with Long was the best way to get something for his story (thus the long wait). On the other hand, perhaps the "bell" issue and Long's investigation prompted Heironimus to make up a story in the first place. Is it feasible that Long told Heironimus to go to Woodard to add credibility to the hoax story?

Whatever the case, Long's book is 476 pages of things people told him. He has a few documents that back up some material, but nothing that proves the film was a hoax.

His story of the alleged "suit" used for the film is a mess. For some reason, he included a claim by costume maker Philip Morris that the suit was one manufactured by his company. This claim conflicts with Heironimus' claim that the suit was made by Patterson from horsehide. Heironimus has also implied (after the book was published) that the suit was made by John Chambers.

There are other very serious discrepancies in the book, too numerous to include here. The glaring omission is that absolutely no hard evidence of any sort is provided to substantiate a hoax. Why anyone would produce a work of this nature without some solid proof is beyond a wonder; it is astounding! I am, of course, stating this from the standpoint that the purpose of the book was to present the truth.

A subsequent television production on Long's findings prompted the following letter from Bob Gimlin to KING TV. The letter was shown and read during a short *Evening Magazine* segment on bigfoot in March 2005.

> I was the only person with Roger Patterson at Bluff Creek on Oct. 20, 1967, when we filmed the creature. I have always believed what I saw was real & not a man in a suit. My belief has been supported over the last 37 years by countless hours of research & scientific studies of the film. I have not profited from the film in any way. Greg Long's book is a crudely written fantasy account of Bob Heironimus' attempt to

Although Long's book did little damage, it did flare a lot of tempers. In essence, what the entire book says is: "The film is a hoax because a lot of people say it is." If issues could be resolved that simply, the sasquatch would have been a scientific reality long before now.

The upside of the book is that Long did provide some good information on Roger Patterson's initiatives. In particular, he got what I consider very positive material from Al DeAtley—providing the first in-depth interview to my knowledge. For all of this, I think we need to say, "thanks Greg."

make a few dollars & enjoy his 15 minutes of fame. More importantly, the book is an ugly character assassination of a man no longer alive to answer the accusations.

6. The Daegling Deception

In early 2005, Dr. David J. Daegling's thoughts and conclusions appeared in his book, *Bigfoot Exposed* (Altamira Press – Rowman & Littlefield Publishers, Inc.). Daegling addresses the entire bigfoot issue in a similar way as Dr. John Napier—generally evasive and throwing in a little humor now and then. In short, Daegling comes to four main conclusions, although, as with Napier, they are "gathered" rather than firmly declared:

1. All footprints are faked.
2. The Patterson/Gimlin film is probably faked. One possibility is a man in skin-tight long underwear with glued-on fur.
3. All sightings are hallucinations or something else other than a sasquatch/bigfoot.
4. Many scientific or other conclusions on the film are wanting or incorrect.

The book is full of incorrect information. Daegling did not do his homework on many things. He did not contact me, even though I am splattered all over the place. He did not even bother to read Dmitri Bayanov's book, *America's Bigfoot: Fact Not Fiction,* the only book devoted to the Patterson/Gimlin film. Above all (in my opinion), rather than using actual photographs, he had an artist copy (draw) photographs. The drawing of a film frame that we know has an anomaly (smudge) due to filming, Daegling points out as a possible hoax indicator.

Here is John Green's review of the book.

BIGFOOT EXPOSED
by Dr. David J. Daegling, Ph.D.,
An Anthropologist Examines America's
Enduring Legend.

No need to pay much attention to this book. There are parts of it that are worth reading, but mostly peripheral to the main issue. The book cover appears to promise skilled dissection by a qualified scientist that disposes of all evidence that sasquatch are real animals, but in that regard the book contains nothing at all.

Leaving out "could be" material like old newspaper stories, Indian traditions, unidentified

I have have a lot of trouble with this book. Even the cover foreshadows its contents—the word "BIGFOOT" is not all there, nor is one's expectations of the book's contents.

Daegling's Foot Fetish

I am not a scientist, and I don't even pretend to be one. However, I have two eyes and a smattering of something called logic.

There are several clear film frames that show what we can say is a "normal" left foot, and one that shows an odd smudged foot (frame 72). René Dahinden pointed this out to me years ago. He said, "you can see that is just the result of the filming, can't you?" I never gave it a second thought.

Dr. Daegling, however, states the following regarding the smudged foot in frame 72 (page 144):

1. The configuration of the Sasquatch heel is unique among primates, or

2. The foot is a prosthetic device worn by a human subject.

He even went to the trouble of having a drawing made of frame 72 to illustrate his point. I am not arguing that the foot is perhaps unique, just that he deliberately used the worst possible frame to make his point.

sounds, smells and hair, mysteriously thrown rocks, and so on, there are three lines of evidence that Dr. Daegling has to explain away: hundreds of casts and photos of footprints, thousands of eye-witness accounts, and one remarkable movie.

As to the footprints, Dr. Daegling has read about them, but there is no indication that he has studied them. Since he is sure that there can be no such animal and that the footprints can easily be faked, he has seen no need to, even when he planned to write about them in a book. What has been reported by the people who actually have investigated such footprints has to be mistaken, because if it were correct the likes of Ray Wallace and Rant Mullens could not have made them, and they have "revealed" that they did.

Eye witnesses? Dr. Daegling goes on at length about the fallibility of human memory. A lot of truth in that, but if its memories were as completely useless as he suggests, the human species could never have survived, let alone written books. He has read the stories of some witnesses, but since there is no such animal and memories are so fallible he has seen no need to talk to any, even when he planned to write a book dismissing all of them as dupes and liars and hallucinators.

Paradoxically, one witness who happened to be a friend of his does seem to have made quite an impression on him, even though hers was a partial on-a-dark-road sighting. He stresses that this lone interview happened to him "not by design," and considering his reaction it seems likely that avoiding talking to people with clear and detailed sightings to describe was absolutely necessary for him to be able to write his book.

It is different regarding the Patterson movie. He has indeed spent time and sought assistance in studying that. Not to prove that it is a hoax, which isn't necessary since there is no such animal, just to try to disprove evidence that what it shows can't be a man in a suit.

Throughout the book there are enough factual errors and ill-founded assumptions to thoroughly mislead anyone who has no other source of information on this subject, but since such a person would not be likely to see this review, it hardly seems worthwhile to deal with them here.

John Green
Harrison Hot Springs,
B.C., Canada
December 23, 2004

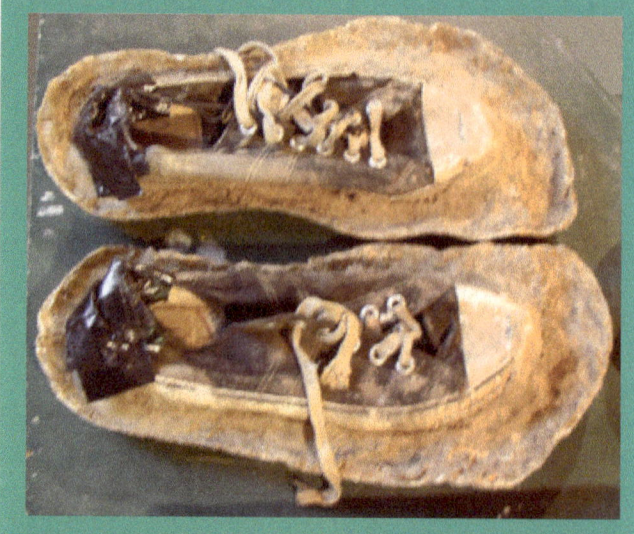

It would appear to me that the best way to fake footprints, if one were so inclined, would be to use actual plaster casts of prints that have been determined to be those of a sasquatch. This set has been fitted with old running shoes to make the job a little more comfortable. However, to make a step of the required length for feet this size would be difficult, and having the needed weight to impress the prints is yet another complication.

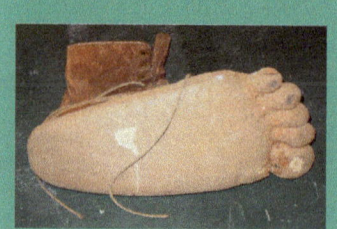

This is a wooden foot that has been fitted with a boot. It appears to be of considerably better craftsmanship than the feet created by Ray Wallace. Professionals, however, are not often fooled by fake prints.

14. Taking Stock

The scientific (forensic) evaluation of the Patterson/Gimlin film performed by NASI *suggests* that Patterson and Gimlin may have seen and filmed a natural bigfoot creature. The circumstantial evidence related to the history of the film is somewhat less convincing. However, even if Mother Teresa took the film, people would still say she faked it. Indeed, some people think the 1969 moon landing was a hoax.

I, along with many others, have certainly "raked the coals" on the film issue and the few "burning embers" that came to the surface were quickly extinguished with undisputable facts.

Credit where credit is due, Greg Long's extensive interviewing does provide additional insights, especially his talk with Al DeAtley. Here DeAtley confirms that he did get the film showing the creature and did get it developed.

Given the creature does exist, the real phenomenon is not so much the creature itself but how it has eluded *capture* in every sense of that word (more photographs, shootings, bones, road kills, and so on). Certainly, in the fifty years since the Jerry Crew discovery, one would think we would have more evidence than we currently have. Personally, I am tired of the "stock answers" to the *capture* questions. I do, however, put a lot of stock in Peter Byrne's suggestion that we are after a creature **that does not want to be caught.** I can also associate with the fact that the total effort to find the creature has not been adequate. In other words, manpower and financial resources have fallen short of that needed for success. As I have stated, there are very few people actually searching for the creature—it is unlikely one will find something if he does not look for it.

Understandably, the allocation of resources (i.e., possible government or research organization financing) relies on *hard* evidence. Other than <u>possible</u> footprints, <u>possible</u> hand prints, <u>possible</u> hair, and <u>possible</u> feces, we have no other hard evidence of bigfoot's existence. The Patterson/Gimlin film and the thousands of sightings, no matter how convincing, are <u>not hard evidence</u>.

We are considerably worse off regarding hard evidence indicating the creature's way of life. It does not appear to consistently build any sort of housing structures, nor does it appear to utilize fire. Some crude structures have been found that cannot be associated with other animals, however, evidence connecting them with bigfoot is very thin. In one instance, a bigfoot was observed stacking rocks after foraging for rodents in a natural rock pile. Just why it stacked the rocks is not

Entrance to the sasquatch exhibit, Vancouver Museum, British Columbia, Canada.

Entrance to the sasquatch exhibit, Museum of Natural History, Pocatello, Idaho, U.S.A.

Not For Lack of Exposure

I have provided two public museum exhibits using my own artifacts/materials and the contributions of many researchers. The first was at the Vancouver Museum, British Columbia, Canada, and ran from June 17, 2004 to February 1, 2005. The second was at the Museum of Natural History, Pocatello, Idaho, USA, and ran from June 16, 2006 to September 11, 2007.

The Vancouver exhibit occupied some 1,800 square feet, and drew over 28,500 visitors. The Pocatello exhibit was smaller, but very skillfully arranged, and highly effective. I was informed that attendance was exceptionally good.

The Patterson/Gimlin material and certainly everything I could obtain of prime importance was included in the exhibits. My primary objective in providing exhibits is to give the sasquatch exposure in hopes of attracting more professional people to consider the sasquatch/bigfoot issue.

Museum virtual tours are at these links:	Vancouver Museum	http://forum.hancockhouse.com/media/van%20museum%20-%20jan%2020%202008.pdf
	Pocatello Museum	http://forum.hancockhouse.com/media/pocatello%20exhibit%20rev%20mar%209.pdf

known. I do not know of another case like this, but it is possible there are other stacked rocks in wilderness areas that may be attributed to bigfoot.

One aspect of the bigfoot issue that is totally overlooked by skeptics and scientists is the vastness of North America's forest regions (particularly in Canada). Most regions are virtually impenetrable. What regions humans have occupied either wholly, partially, or incidentally is insignificant compared to the regions they have not occupied or even explored at ground level. In some ways, this situation is the same as the claim that the vastness of the universe precludes the idea that human beings are the only "intelligent" beings in existence. In other words, aliens of some sort are probably a given. By the same token, there can be no doubt that a creature such as bigfoot could exist in North America with only marginal notice by humans. Here I will add that if the odd one did not inadvertently cross paths with humans, we wouldn't even know they were there.

Be that as it may, ironically the whole bigfoot issue has resulted in a "Catch-22" situation. Hard evidence is needed to seek resources, but resources are needed to seek hard evidence. Many people contend that current knowledge and the potential prize justifies the gamble. In other words, what we have as evidence, and the possibility of finding a bigfoot, should be enough incentive for a major undertaking by perhaps the Smithsonian Institution or the National Geographic Society. Such contentions are reasonably well founded. Bigfoot is the only phenomenon I know that appears to continually leave calling cards—sightings are one thing, but footprints are a totally different story.

A Final Word

The bigfoot filmed by Patterson and Gimlin has probably crossed over to the other side by this time. Whether she was ever seen again will never really be known. Other people certainly reported sightings in the Bluff Creek area after 1967, so it is entirely possible she was seen again.

Remarkably, the great flurry of bigfoot activity in late 1967 was also the first and last time anything like that occurred. Certainly with all of the prints found on Blue Creek Mountain, and then the filming of a bigfoot, everyone expected to resolve the issue very quickly. But that did not happen. Over forty years have slipped by and we have yet to get a photograph, movie, or video of the creature as good as that taken by Roger Patterson and Bob Gimlin.

Did you ever notice that there is no difference between a sunrise and a sunset?

BIBLIOGRAPHY

SUBJECT BOOKS, ESSAYS AND PAPERS

Attenborough, David and Molly Cox, 1975. *Fabulous Animals.* London, England: BBC.
Daigling, David, 2004. *Bigfoot Exposed.* Walnut Creek, California: AltaMira Press.
Donskoy, Dmitri D., 1971. Qualitative Biomechanical Analysis of the Walk of the Creature in the Patterson Film. Moscow, Russia: (Paper).
Glickman, J., 1998. *Toward a Resolution of the Bigfoot Phenomenon.* Hood River, Oregon: North American Science Institute.
Krantz, Grover S., 1992. *Big Footprints: A Scientific Inquiry into the Reality of Sasquatch.* Boulder, Colorado: Johnson Printing Co.
Krantz, Grover S., 1999. *Bigfoot/Sasquatch Evidence.* Surrey, B.C.: Hancock House Publishers.
Long, Greg, 2004. *The Making of Bigfoot.* Amherst, New York: Prometheus Books.
Markotic, Vladimir, 1984. *The Sasquatch and Other Unknown Hominids.* Calgary Alberta: Western Publishing.
Meldrum, D. Jeffrey, 2006. *Sasquatch: Legend Meets Science.* New York, New York: Tom Doherty Associates.
Murphy, Christopher L., with John Green and Thomas Steenburg, 2004. *Meet the Sasquatch.* Surrey, B.C.: Hancock House Publishers.
Napier, John, 1972. *Bigfoot: Startling Evidence of Another Form of Life on Earth.* New York, New York: Berkley Publishing.
Patterson, Roger, 1966. *Do Abominable Snowmen of America Really Exist?* Yakima, Washington: Franklin Press. Reprinted, 1997, New Westminster, B.C.:Pyramid Publications.
Patterson, Roger, 1966 or 1967. *Do Abominable Snowmen of America Really Exist? (Second edition).* Yakima Washington: Northwest Research Association.
Patterson, Roger, with supplement by Christopher L. Murphy, 2005. *The Bigfoot Film Controversy.* Surrey, B.C.: Hancock House Publishers (reprint of Patterson's book, *Do Abominable Snowmen of America Really Exist?*).
Patterson, Roger, 1968. *Bigfoot - Volume 1.* Tampico, Yakima County, Washington: Self-published.
Perez, Daniel Edward, 1992. *BigfooTimes: Bigfoot at Bluff Creek.* Norwalk, California: Self-published. Reprinted in 2003 with a bibliography.

NEWSPAPER /MAGAZINE ARTICLES AND LETTERS

Harrison, George H., 1970. "On the Trail of Bigfoot." Milwaukee, Wisconsin: *National Wildlife* magazine. October – November.
McConnel, Rob, June 1996. *X Chronicles* article. Letter from Harry Kemball dated May 14, 1996.
Sanderson, Ivan T., 1959. "The Strange Story of America's Abominable Snowman": *True* magazine, December.
Sanderson, Ivan T., 1968. "First Photos of 'Bigfoot,' California's Legendary Abominable Snowman": *Argosy* magazine, February.
Wasson David, 1999. "Bigfoot Unzipped, Man Claims It Was Him in a Suit!" Yakima, Washington: *Yakima Herald–Republic.* January 30.
(no by-line), 1964. "Allprints Photo Adds New Unique Film Processing Unit," Mansfield, Ohio: *News Journal.* November 16.
(no by line), 1967. (Reporter believed to be Al Tostado). "Mrs. Bigfoot Is Filmed." Eureka, California: *The Times-Standard.* October 21.

TELEVISION DOCUMENTARIES

The World's Greatest Hoaxes Revealed (1998), Fox.
Sasquatch: Legend Meets Science (2003), White Wolf Entertainment, Inc.

LINKS – PAGE NUMBERS

Drawing for the film site model, 22
Creature/Female comparison, 24
Film site cast in 3D, 33
Article by Al Tostado, 41
Study by C.L. Murphy, 47
Patterson's booklet, 55
Court/Legal documents – P/G film, 56
National Wildlife Article, 67
Bourtsev sculpture, 82

The Bigfoot Research Project brochure, 83
Marlon Davis' findings, 85
Film Resolution, 88
Bobbie Short's web site (Bigfoot Encounters), 89
Heironimus Comparison, 93
Museum virtual tours, 97
Errata Link, 105
Hancock House Publishers web site, 106

PHOTOGRAPH CREDITS/COPYRIGHTS

Page	Description	Credit/Copyright/Source
Front cover	Film strip	Author's collection
Frontispiece		M. Rugg
7	Map	Google Earth
8	Murphy/Dahinden	C.L. Murphy
8	Dahinden/others/portable building	C.L. Murphy
8	Dahinden	C.L. Murphy
9	Murphy	Gerry Matthews
9	Group	C.L. Murphy
10	Books	C.L. Murphy
11	Book Cover	C.L. Murphy
12	John Green	J. Green
12	Map	Y. Leclerc/C.L. Murphy
13	Roger Patterson	J. Green
13	Robert Gimlin	C.L. Murphy
13	Book Cover	Franklin Press
13	Camera ad	Eastman Kodak
14	Jerry Crew	*Humboldt Times*
14	Cast	C.L. Murphy
15	Fake Casts	C.L. Murphy
15	Satellite Photos showing roads	Google Earth - Tele Atlas/DigitalGlobe
16	Map	J. Green
17	Movie Camera (two photos)	C.L. Murphy
17	Film strip	C.L. Murphy
18	Frame 61 artwork	B. Bannon
19	Frame 307 Artwork	B. Bannon
20	Frame 323 Artwork	B. Bannon
21	Frame 343 Artwork	B. Bannon
22	File 352 Artwork	C.L. Murphy
22	Film Site Model	C.L. Murphy
23	Frame 364 Artwork	B. Bannon
24	Bluff Creek Soil	C.L. Murphy
24	Frame 352/Female Form	E&M Dahinden - Public Domain
25	Bigfoot/Rifle Scope	Y. Leclerc/C.L. Murphy
26	Wood Fragment (5 photos)	C.L. Murphy
27	Film Site Model - 3 angles	C.L. Murphy
28	Footprints - 3 photos	R.L. Laverty
29	Bigfoot Diagram	J. Green
30	Frame 352 with measurements	E&M Dahinden - Public Domain
30	Map - R. Titmus	R. Titmus/Author's collection
30	Film Site Model	C.L. Murphy
30	Frame 352/Film Site Model diagram	E&M Dahinden - Public Domain/C.L. Murphy
31	Frame 352	E&M Dahinden - Public Domain
31	Film Site with M. Hodgson	P. Byrne
32	Film Site Satellite Image	Google Earth - Tele Atlas/DigitalGlobe
32	Film Site (3 photos)	C.L. Murphy
33	Film site Casts/Human foot cast	C.L. Murphy
33	Casts - Titmus (4 photos)	C.L. Murphy
34	Al Hodgson	D. Perez
34	Willow Creek in 1960	Author's collection
35	Murray Field Airport	C.L. Murphy

35	Murray Field Airport Office	C.L. Murphy
35	Plane	C.L. Murphy
35	McClarin and Murphy	C.L. Murphy
36	Syl McCoy	J. Green
36	Bayanov and McClarin	C.L. Murphy
39	Don Abbott	J. Green
39	White Lady	J. Green
40	Hodgson's Department Store	A. Hodgson
40	Willow Creek–China Flat Museum	C.L. Murphy
40	Willow Creek, 2003	C.L. Murphy
41	The Times-Standard Building	D. Perez
41	*The Times-Standard* Banner	*The Times-Standard*
42	Sasquatch	Y. Leclerc.
43	Woods - Frame 352	E&M Dahinden - Public Domain
43	Frame 352 with insert	E&M Dahinden - Public Domain
44	Note to Dahinden	D. Perez
45	Newspaper	*News Journal* - McClean collection
46	Book Cover	BBC/D. Attenborough
47	Sasquatch Study	C.L. Murphy
47	Sasquatch Head - Left	R. Menzies
47	Sasquatch Head - Right	P. Travers
48	Patterson with pony	J. Green
48	Dahinden and Patterson	J. Green
48	Gimlin's home	C.L. Murphy
49	Patterson and Jensen	J. Green
49	Satellite photo	Google Earth - Digital Globe
50	Robert Gimlin portrait	C.L. Murphy
50	Gimlin at podium	C.L. Murphy
51	Newspaper ad	Author's collection
52	Note from Gimlin	C. Crook
52	Movie Notice	J. Green
53	Agreement for book	G. Koelling
53	Patterson Book covers (4)	E&M Dahinden/Hancock House
54	Flower	C.L. Murphy
55	Certificate - Abominable Snowman Club	E&M Dahinden/Hancock House
55	Certificate - Northwest Research	Author's collection
56	Court documents	C.L. Murphy
56	Les Patterson's business card	Author's collection
57	Dahinden and Patterson	J. Green
58	Gimlin and Murphy	T. Steenburg
58	Book and Gavel	C.L. Murphy
59	University entrance pillar	C.L. Murphy
59	Satellite photo	Google Earth - Europa Technologies
60	Bob Titmus	J. Green
60	List of Names	J. Green
61	John Green	C.L. Murphy
61	John Green's House	C.L Murphy
62	Book cover	Hancock House
62	Krantz and Donskoy	G. Krantz
63	Museum of Anthropology	C.L. Murphy
63	Pole carvings (2 photos)	C.L. Murphy
64.	Don Abbott	J. Green
65	Hotel Georgia (3 photos)	C.L. Murphy

65	Heritage Plaque	C.L. Murphy
66	David Hancock	C.L. Murphy
66	Groenheyde, Murphy, Hancock	C.L. Murphy
66	Book Cover	Hancock House
67	Dr. Clifford Carl	J. Green
67	Polygraph	Author's collection (Public Domain)
69	Frank Beebe	David Hancock
69	Royal Provincial Museum	Ryan Bushby
71	Book Cover - Dr. J. Napier.	Berkley Publishing
72	Universal Studios - Fountain	Anthony Giorgio (See Note below)
72	Universal Studios - Red Carpet	Sarge Baldy (See Note below)
72	Bronx Zoo	Elgen Werl (See Note below)
73	Smithsonian Institution	C.L. Murphy
77	*Argosy* Cover	*Argosy* magazine - Author's Collection
77	Patterson and Gimlin	P. Patterson/Jennings (first name not known)
80	Dahinden in Red Square	D. Bayanov
80	Dahinden with group	D. Bayanov
81	Bourtsev and Bayanov	D. Perez
81	Dr. D. Donskoy	Igor Bourtsev
81	Bayanov and Goodall	D. Bayanov
82	Bourtsev statue (4 photos)	C.L. Murphy
83	Dahinden, Byrne, Murphy	P. Byrne
83	Bigfoot Research Project Brochure (2 photos)	Boston Academy of Applied Science
84	Bigfoot Research Project illustration	Boston Academy of Applied Science
84	Dahinden and Byrne - Dahinden's place	C.L. Murphy
84	Byrne with daughter and friend	C.L. Murphy
84	Lund and Byrne	C.L. Murphy
84	Byrne and Dahinden at Carson	C.L. Murphy
85	Sasquatch Legends Meet Science DVD	White Wolf Entertainment
85	Sasquatch Legends Meets Science book cover	White Wolf Entertainment/Dr. J. Meldrum
86	Bigfoot silhouette	C.L. Murphy
86	Bigfoot silhouette setup	C.L. Murphy
87	Height diagrams (2 images)	B. Bannon/C.L. Murphy
88	Frame 352 with dot	E&M Dahinden (Public Domain)
88	Drums	Author's file
89	Bobbie Short	Bobbie Short
89	John Chambers,	Bobbie Short
89	Patterson sculpture	C.L. Murphy
93	Heironimus/bigfoot diagram	R. Noll/B. Bannon
94	Book cover	Promethus Books
95	Book cover,	Altamira Press–Roman & Littlefield Inc.
96	Fake Feet (3 photos)	C.L. Murphy
97	Museum entrance - Van. Museum	C.L. Murphy
97	Museum entrance - Idaho Museum of Nat. His.	C.L. Murphy
98	Bigfoot Crossing,	C.L. Murphy
98	Sunrise	C.L. Murphy
Back Cover – Film site model		C.L. Murphy
Back Cover – Sasquatch artwork (frame 364)		B. Bannon
Back Cover – Footprint casts		C.L. Murphy

NOTE: These images are from Wikipedia Encyclopedia and are used used with permission under the Creative Commons License. Please refer to the Wikipedia website for details.

GENERAL INDEX

Abbott, Don, (Obituary), 68
Abbott, Don, 12, 14, 34, 35, 39, 43, 44, 59, 60, 63, 64, 67, 68, 69, 70
Abominable Snowman – Legend Come to Life, 75
Abominable Snowmen Club of America (certificate), 55
Abominable Snowmen Club of America, 49
Allen, Tom, 75
Allied Holding Ltd., 65
Allprints Photo, Inc., 45, 46
America's Bigfoot: Fact Not Fiction, 36, 95
American Museum of Natural History, 72
American National Enterprises, 47, 91, 92
Arcata (California), 37, 60
Argosy magazine, 72, 73, 75, 77, 78, 80, 81
Associated Press, 92, 93
Atlanta (Georgia), 72, 73
Attenborough, Sir David, 46
Aviation Institute (Moscow, Russia), 82

B.C. Provincial Museum, 59
Bannon, Brenden, 18
Bayanov, Dmitri, 8, 9, 22, 36, 80, 81, 82, 95
BBC, 46
Beaverton (Oregon), 78
Beck, Fred, 8
Beebe, Frank, 60, 63, 64, 67, 68, 69, 70
Bella Coola (British Columbia), 42
Bering Sea, 76
Bering Strait, 62, 76
Bible, 74
Bigfoot: Monster, Myth or Man, 74
Bigfoot (book), 68, 77
Bigfoot (Volume 1), 55
Bigfoot Encounters (website), 89
Bigfoot Enterprises, 50
Bigfoot Exposed, 95
Bigfoot Film Controversy, 7, 53 (cover), 54
Bigfoot Research Project, The, 83, 84, 90
Bigfoot Symposium - Willow Creek, 35
Bigfoot Times (Newsletter), 37
Bigfoot/Sasquatch Evidence, 51, 62
Bigfoot Bulletin (Patterson), 76
Bing Thom & Associates, 65
Block and Company, 15
Blue Creek Mountain (California), 12, 14, 15, 39, 64, 98
Bluff Creek (California), 12, 13, 14, 15, 16, 17, 18, 19, 21, 24, 32, 34, 35, 36, 38, 39, 40, 41, 48, 50, 52, 53, 57, 78, 79, 89, 90, 91, 94, 98
Boeing Laboratories (Seattle), 44
Bonney, Bruce, 74
Bothell (Washington), 92
Bourtsev, Alexandra, 81
Bourtsev, Igor, 22, 80, 81, 82
Breazele, Bob, 15
British Columbia Provincial Museum, 12, 14, 34, 68
Bronx Zoo (New York), 72, 74
Bryan, Allan, Dr., 75
Byrne, Peter, 16, 17, 25, 26, 31, 74, 83, 84, 85, 87, 90, 91, 97
Byrne, Rara, 84

CanaWest (film facility), 90, 91
Carl, Clifford, Dr., 67, 70
Chambers, John, 89, 90, 94,
Chicago (Illinois), 45
Chico (horse), 38
Circumstantial Evidence - The Patterson/Gimlin Film, 8, 11, 93
Cole, Nat King, 65
Columbus (Ohio), 45
Cook, Joedy, 66
Cranbrook (British Columbia), 90
Crew, Jerry, 12, 14, 15, 16, 97
Crook, Cliff, 52, 56, 92
Crowe, Ray, 84

Daegling, David J., Dr. 95, 96
Dahinden, Erik, 56
Dahinden, Martin, 56
Dahinden, Rene, 8, 9, 12, 14, 19, 25, 26, 31, 35, 36, 40, 44, 45, 46, 47, 48, 49, 50, 52, 53, 54, 56, 57, 59, 60, 67, 69, 71, 74, 80, 81, 82, 83, 84, 87, 91, 93, 95
Darwin Museum (Moscow, Russia), 81
Davis, Marlon K., 19, 36, 85
DeAtley, Al, Jr., 9, 34, 35, 36, 37, 38, 43, 44, 45, 46, 47, 50, 51, 52, 57, 58, 72, 89, 94, 97
DeAtley, Iva, 57
Digest magazine, 42
Director of Management Operations – U.S. Department of the Interior, 72
Disney Studios (Hollywood, California), 73, 74
Do Abominable Snowmen of America Really Exist? 7, 8, 12, 13, 41, 48, 53 (book covers), 54, 75

Donskoy, Dmitri, Dr., 62, 81
Emory University (New York), 73
Eureka (California) 35, 37, 39, 40, 43
Evening Magazine (television show), 94
Fabulous Animals, 46
Fahrenbach, Henner, Dr., 8, 22, 66, 79, 88, 92
Ford Motor Company, 52
Fossey, Dian, 79
Franklin Press, 7, 53
Freeman, Paul, 28

Garrow, R.T., 65
George the Fifth Pub, 65
Georgia Hotel (Vancouver, B.C.), 59, 65
Georgian Club (Vancouver, B.C.), 65
Gimlin, Robert (Bob), 7, 8, 9, 10, 12, 13, 14, 15, 17, 18, 19, 20, 21, 22, 23, 24, 25, 26, 27, 28, 34, 35, 36, 37, 38, 39, 40, 41, 42, 43, 47, 48, 50, 51, 52, 54, 55, 56, 57, 58, 59, 60, 66, 69, 72, 77, 78, 79, 84, 85, 87, 89, 91, 93, 97, 94, 98
Glickman, Jeff, 7, 21, 22, 26, 58, 61, 83, 84, 85, 88
Golden Eagle Productions, 90
Goodall, Jane, Dr. 81
Graham, John., Sr., 65
Green, Beverly, 60
Green, John, 7, 8, 9, 12, 14, 16, 20, 22, 29, 35, 36, 39, 44, 46, 48, 49, 50, 52, 59, 60, 61, 66, 67, 69, 70, 71, 73, 74, 79, 87, 91, 92, 95, 96
Grieve, Donald, Dr., 22, 80, 81
Groenheyde, Rick, 66
Guiget, Charles, 60

Hajicek, Doug, 66
Hancock, David, 65, 66
Harrison Hot Springs (British Columbia), 96
Harrison, George H., 67
Harvey, Steven, 64
Healy, Tony, 38
Heironimus, Bob, 38, 67, 93, 94, 95
Hill, Osman, Dr., 73, 77
Himalayas, 71
Hodgson, Al, 12, 14, 22, 34, 35, 37, 38, 39, 40, 44
Hodgson, Francis. 38
Hodgson, Michael, 31
Hollywood (California), 72, 74, 77, 89

Hooker, Charlie, Mr. and Mrs., 13, 17
Hotel Georgia (Vancouver, B.C.), 59, 65
Houck, W.J., Professor, 60, 65
Humboldt County (California), 41, 52
Humboldt State College (Arcata, California, 60, 65
Humboldt Times, 14
Hunt, J.B., 52

I Fought the Apemen of Mt. St. Helens, 8
International Society of Cryptozoology, 81

J.B. Hunt Transport, Inc., 52
Jensen, Dennis, 49

Kemball, Harry, 89, 90, 91
Kerr, Ray, 15
Kiernan, Kenneth, 60, 67, 69, 70
King Edward VIII, 65
KING TV, 94
Knights, Roger, 45
Kodak (ad), 13
Kodak Technicolor Laboratory (Seattle), 36, 44
Koelling, Glen, 49, 53, 54
Koffmann, Marie-Jeanne, Dr., 80
Kopas, Cliff, 42
Krantz, Grover, Dr., 19, 22, 25, 51, 62 77, 87

Leclerc, Yvon, 22, 25, 42, 47, 66
Life magazine, 25, 72, 74
London (England), 80
Long, Greg, 7, 38, 54, 57, 58, 90, 93, 94, 97
Loudon, Peter, 44
Louse Camp (California), 12
Lower Trinity Ranger Station (California), 22, 35, 37
Lund, Larry, 84

Majors, Kenneth C., 45, 46
Making of Bigfoot, The, 38, 54, 57, 90, 93
Mansfield (Ohio), 46
Markotic, Vladimir, Dr., 75, 77, 81
Martins Ferry (bridge - California), 37
Marx, Ivan, 28
Matthews, Gerry, 9
McClarin, Jim, 35, 36
McCoy, Syl, 12, 14, 22, 35, 39, 40
McTaggart-Cowan, Ian, Dr., 60, 61, 65
Meet the Sasquatch, 47, 66

103

Meldrum, Jeff, Dr., 66, 79, 85, 87
Memorial Day video, 28
Menzies, RobRoy, 47
Mondor, Patricia (Patricia Patterson), 13
Montagna, William, Dr., 78
Morgan, Robert, 66
Morris, Philip, 94
Moscow (Russia), 62, 80
Mother Teresa, 97
Mount St. Helens (Washington), 12, 38
Murphy File, The, 54
Murphy, Christopher L. 9, 47, 92
Murphy, Daniel, 8
Murray Field (Airport - California), 35, 38
Museum of Anthropology (British Columbia), 63
Museum of Natural History (New York), 85
Museum of Natural History (Pocatello, Idaho), 97
Mystical Biped, The (article), 81

Napier, John, Dr., 16, 67, 70, 73, 75, 76, 77, 78, 95
NASI Report, 82, 84, 85, 88, 97
National Geographic Society, 73, 75, 77, 98
National Wildlife magazine 67, 80
New York (New York), 72, 74
New York Times, The, 73
News Journal (Ohio), 45
Newsman's Club (Vancouver, B.C.), 65
Noll, Richard, 7, 66, 87
North American Aviation (Columbus, Ohio), 45
North American Science Institute (NASI), 7, 22, 83, 88
North Vancouver (British Columbia), 90
Northwest Research Association, 17, 49, 53, 73
Northwest Research Association (certificate), 55
Notice Creek (California), 12, 41

Olson, Ron, 54
Onion Mountain (California), 35
Orleans (California), 36
Oroville (California), 49

Pacific Northwest Expedition, 42
Palo Alto (California), 44
Pasco (Washington), 52
Paterson, T.W., 42
Patterson, Brad, 13
Patterson, Clint, 13
Patterson, Glen, 52
Patterson, Les, 52, 56

Patterson, Lorn, 52
Patterson, Patricia (Mrs. Patterson), 8, 12, 51, 52, 56, 91
Patterson, Roger, 7, 9, 10, 12, 13, 14, 15, 16, 17, 18, 19, 20, 21, 22, 23, 24, 25, 26, 27, 28, 30, 33, 34, 35, 36, 37, 38, 39, 40, 41, 42, 43, 44, 45, 46, 47, 48, 49, 50, 51, 52, 53, 54, 55, 56, 57, 58, 59, 60, 64, 65, 66, 67, 69, 72, 73, 74, 75, 76, 77, 78, 79, 80, 82, 85, 86, 87, 89, 90, 91, 93, 94, 98
Patterson, Sabrena, 13
Patty (creature in Patterson/Gimlin film), 36, 43, 65, 88
Peanuts (Welsh Pony), 23, 38
Perez, Daniel, 8, 21, 32, 37, 38, 44, 66
Philadelphia (Pennsylvania), 45
Planet of the Apes, 89, 90
Post, The, 73
Presley, Elvis, 65
Prohaska, Janos, 74
Prop-Lock (invention), 48
Provincial Museum (British Columbia), 59, 67, 69, 70
Pyramid Publications, 8, 54

Queen Elizabeth II, 69

Radford, George W., Jr., (Mr. and Mrs.), 49, 50, 56
Ray Littmann & Associates, 65
Reader's Digest, 80
Real West magazine, 42
Redwoods video, 28
Reinke, Clyde, 91
Relic Hominoid Research Seminar (Moscow, Russia), 81
Richmond (British Columbia), 25
Rochester, New York, 44
Roe, William, 23
Rolling Stones, 65
Romney, Jerry, 91
Royal British Columbia Museum, 69
Russian Central Scientific Research Institute of Prosthetics and Artificial Limb Construction, 80

Samarkand, Ukbekistan (USSR/Russia), 82
San Francisco (California), 44
Sanderson, Ivan T., 72, 73, 74, 75, 81
Sarmiento, Esteban, Dr., 21, 85
Sasquatch and Other Unknown Hominoids, 77, 81
Sasquatch Returns, The (article), 66
Sasquatch/Bigfoot, 8

Sasquatch: Legend Meets Science (book), 85
Sasquatch: Legend Meets Science (documentary), 85
Schaller, George, Dr., 79
Schuchman, Cheryl, 38
Schuchman, John, 38
Science and Religion magazine, 81
Seattle (Washington), 17, 36, 44
Shapiro, Harry, Dr., 74
Shaw and Sons Chapel, 54
Sheets, Dale, 73
Sheppard's Drive-In Camera Store, 13
Shooting the Bigfoot (documentary), 57
Short, Bobbie, 89, 93
Six Rivers National Forest (California), 12, 32
Slate Creek (California), 38
Slick,Tom, 16, 42
Smithson, James, 73
Smithsonian Institution, 73, 74, 75, 76, 78, 79, 98
Smolin research group, 81
Snow Walker video, 28
Steenburg, Thomas, 8, 9, 20, 37, 38, 66, 77, 87
Strange Story of America's Abominable Snowman (article), 75

Tampico (Washington), 13
Thorington, Richard, Jr., Dr., 78
Times Colonist The, 45, 68
Times Standard, The, 17, 21, 22, 35, 38, 40, 41, 44, 91
Titmus, Bob, 15, 27, 30, 33, 42, 60, 69, 87
Toledo (Washington), 16
Tostado, Al, 41
Toward a Resolution of the Bigfoot Phenomenon, 7, 83
Trail Blazers Research Institute, Inc., 49, 53
Travers, Peter, 47
True magazine, 75

U.S. Coast and Geodetic Survey, 75
U.S. Department of Commerce, 75
U.S. Department of the Interior, 75
UBC (University of British Columbia), 59, 65
Udall, Stewart, 75
Universal Studios (Hollywood, California), 72, 73
University of Alberta, 75
University of British Columbia (UBC), 35, 37, 46, 59, 79

University of Calgary, 75
University of Victoria (British Columbia), 67
USSR Committee on Cinematography, 80
USSR Institute of Physical Culture, 81
Van Gelder, Richard, Dr., 74
Vancouver (British Columbia), 35, 59, 65, 97
Vancouver Museum (British Columbia), 65, 66, 97
Victoria (British Columbia), 44, 59, 64, 70
Victoria National History Society, 67
Victoria *Times-Colonist*, 45, 68

Wallace, Ray, 12, 15, 16, 89, 96
Walls (South Dakota), 13
Washington, D.C., 73
Wasson, David, 93
Wayne, John, 65
Webb, Mallory, 84
Webster, Jack, 59, 60
Weekend Magazine, 66
West Hills Memorial Park (Yakima), 54
West magazine, 80
Western Publishers (Calgary, Alberta), 77
White Lady (tracking dog), 39
White Wolf Entertainment, 85
Willow Creek–China Flat Museum, 15, 35, 40
Willow Creek (California), 9, 12, 34, 35, 37, 40, 41, 44
Willow Creek Bigfoot Symposium, 36, 58
Wood, N.O., 75
Woodard, Barry M., 93, 94
World's Greatest Hoaxes, The (documentary), 46, 91
Wraight, A. Dr., 75

X Chronicles, 89, 90, 91

Yakima (City), 13
Yakima (Washington), 17, 35, 36, 37, 38, 43, 44, 49, 52, 93
Yakima County (Washington), 13, 50, 93
Yakima Herald Republic, 49, 93
Yakima, 54
Yerkes Regional Primate Research Center (Emory University, New York), 73
Yeti, 71

Zana, 82
Zilla (Washington), 93
Znanie-Sila magazine, 81

Best of Sasquatch Bigfoot
John Green
0-88839-546-9
8½ x 11, sc, 144 pages

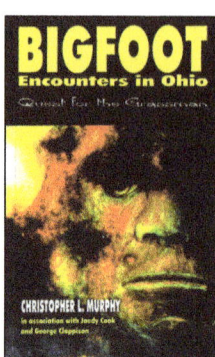
Bigfoot Encounters in Ohio
C. Murphy, J. Cook, G. Clappison
0-88839-607-4
5½ x 8½, sc, 152 pages

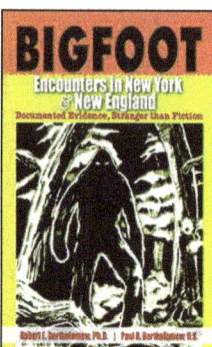
Bigfoot Encounters in New York & New England
Robert Bartholomew
Paul Bartholomew
978-0-88839-652-5
5½ x 8½, sc, 176 pages

Bigfoot Film Controversy
Roger Patterson, Christopher Murphy
0-88839-581-7
5½ x 8½, sc, 240 pages

Bigfoot Film Journal
Christopher Murphy
0-88839-658-7
8½ x 11, sc, 106 pages

Bigfoot Sasquatch Evidence
Dr. Grover S. Krantz
0-88839-447-0
5½ x 8½, sc, 348 pages

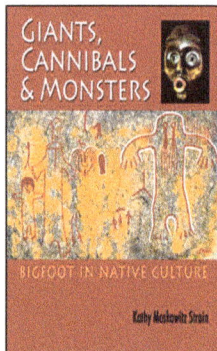
Giants, Cannibals & Monsters: Bigfoot in Native Culture
Kathy Moskowitz Strain
0-88839-650-3
8½ x 11, sc, 288 pages

Hoopa Project
David Paulides
0-88839-653-2
5½ x 8½, sc, 336 pages

In Search of Giants
Thomas Steenburg
0-88839-446-2
5½ x 8½, sc, 256 pages

The Locals
Thom Powell
0-88839-552-3
5½ x 8½, sc, 272 pages

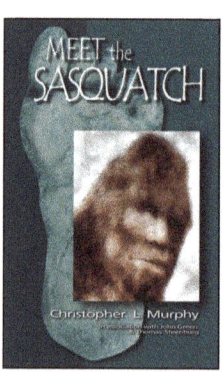
Meet the Sasquatch
Christopher Murphy,
John Green,
Thomas Steenburg
0-88839-574-4
8½ x 11, hc, 240 pages

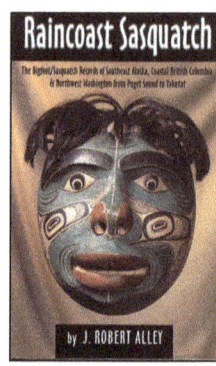
Raincoast Sasquatch
J. Robert Alley
978-0-88839-508-5
5½ x 8½, sc, 360 pages

Sasquatch: The Apes Among Us
John Green
0-88839-123-4
5½ x 8½, sc, 492 pages

Sasquatch Bigfoot
Thomas Steenburg
0-88839-312-1
5½ x 8½, sc, 128 pages

Sasquatch/Bigfoot and the Mystery of the Wildman
Jean-Paul Debenat
978-0-88839-685-3
5½ x 8½, sc, 428 pages

Strange Northwest
Chris Bader
0-88839-359-8
5½ x 8½, sc, 144 pages

Tribal Bigfoot
David Paulides
978-0-88839-687-7
5½ x 8½, sc, 336 pages

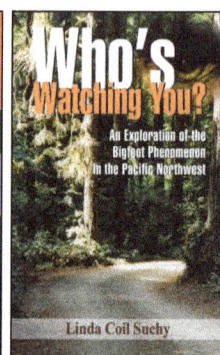
Who's Watching You?
Linda Coil Suchy
978-0-88839-664.8
5½ x 8½, sc, 408 pages

View all **HANCOCK HOUSE** *titles at* **www.hancockhouse.com**